Weather Economics

UNIVERSITY COLLEGE OF WALES, ABERYSTWYTH

MEMORANDUM NO 11 (1968)

Weather Economics

BASED ON PAPERS AND DISCUSSIONS AT A SYMPOSIUM HELD AT THE WELSH PLANT BREEDING STATION NEAR ABERYSTWYTH ON MARCH 13TH, 1968

Edited by

JAMES A. TAYLOR

Senior Lecturer in Geography at the University College of Wales, Aberystwyth

PERGAMON PRESS

Oxford · London · Edinburgh · New York
Toronto · Sydney · Paris · Braunschweig

Pergamon Press Ltd., Headington Hill Hall, Oxford
4 & 5 Fitzroy Square, London W.1
Pergamon Press (Scotland) Ltd., 2 & 3 Teviot Place, Edinburgh 1
Pergamon Press Inc., Maxwell House, Fairview Park, Elmsford, New York 10523
Pergamon of Canada Ltd., 207 Queen's Quay West, Toronto 1
Pergamon Press (Aust.) Pty. Ltd., 19a Boundary Street,
Rushcutters Bay, N.S.W. 2011, Australia
Pergamon Press S.A.R.L., 24 rue des Écoles, Paris 5e
Vieweg & Sohn GmbH, Burgplatz 1, Braunschweig

First edition 1970

Library of Congress Catalog Card No. 75-97951

Printed in Great Britain by A. Wheaton & Co., Exeter

08 006766 2

Contents

Editor's Acknowledgements xi

Aberystwyth Memoranda xiii

Introduction 1

Chapter 1. The Cost of British Weather 5
by JAMES A. TAYLOR

Chapter 2. The Effect of the Weather on Farm Organization and Farm Management 11
by J. M. STANSFIELD

Chapter 3. Weather and Machinery Work-days 17
by C. V. SMITH

Chapter 4. Basic Frost, Irrigation and Degree-day Data for Planning Purposes 27
by W. H. HOGG

Chapter 5. Variations in the Marginal Value of Agricultural Labour Due to Weather Factors 45
by W. J. TAGGART

Chapter 6. A Note on the Areal Patterns in the Value of Early Potato Production in South-west Wales, 1967 51
by J. G. TYRRELL

Chapter 7. Weather and Risk in Forestry 67
by P. A. WARDLE

Chapter 8. The Use of Cost-benefit Studies in the Interpretation of Probability Forecasts for Agriculture and Industry: an Operational Example 83
by E. T. STRINGER

Chapter 9. Edited Report of the Discussions Held at Symposium XI, 1968 93
by JAMES A. TAYLOR

Chapter 10. Economic Postscript 107
by G. N. Rubra
Author Index 121
Subject Index 123

List of Figures

Fig. 1. Table showing input, output, and efficiency measures for predominantly arable farms (with more than 50 per cent tillage) 12

Fig. 2. Cumulative percentage frequency of dry spells in May: Nottingham (1916–50) 21

Fig. 3. Work criteria applied to spring cultivations on "medium" land in the east Midlands 23

Fig. 4. Cumulative percentage frequencies of total work-days available for spring cultivations (February, March, April) for "heavy", "medium" and "light" soils in the east Midlands 24

Fig. 5. The relationship between frost frequencies (years in 10) and weekly mean cumulative temperatures at Long Ashton, near Bristol 29

Fig. 6. Table showing the results of a frost survey on a Somerset farm (1963–5) 30

Fig. 7. Table showing the variation in regional frost risk in Somerset 31

Fig. 8. Table showing the number of occasions with minimum screen temperatures $\leqslant 32°F$ ($0 \cdot 0°C$) on one or more nights at Long Ashton, near Bristol, between the stated dates and the end of the frost season 32

Fig. 9. Table showing irrigation-needs at Oxford, 1949 33

Fig. 10. Total irrigation-need over 20 years (in). Period: April–September. S.M.D. 1 in. 35

Fig. 11. Irrigation-need for driest year in 20 (in.). Period: April–September. S.M.D. 1 in. 36

Fig. 12. Total irrigation-need over 20 years (in.). Period: April–June. S.M.D. 1 in. 37

FIG. 13. Irrigation-need for driest year in 20 (in.). Period: April–June. S.M.D. 1 in. 38

FIG. 14. Table showing irrigation-needs, Shrewsbury, April–September. S.M.D. 1 in. 39

FIG. 15. Table showing irrigation-needs, Pembrokeshire coast, April–June. S.M.D. 1 in. 39

FIG. 16. Table showing averages of monthly degree-days below stated bases for Penzance, Cornwall 40

FIG. 17. Data for an example of computation of degree-days for apportionment of fuel costs 42

FIG. 18. Crops considered for inclusion in the farm plans. 46

FIG. 19. Comparison of proportions of time suitable for outdoor work (periods of two weeks) 47

FIG. 20. Comparison of proportions of time suitable for outdoor work (seasonal periods) 48

FIG. 21. Effect of labour availability variation on crop distribution selected 49

FIG. 22. Comparative summary of enterprise distribution, profitability, and marginal value of labour 50

FIG. 23. Weather factors in the growth stages of the potato plant. (Adapted from McQuigg and Doll) 53

FIG. 24. *Home Guard* prices, 1967 54

FIG. 25. West Pembrokeshire: early potato acreages by parish (percentages of total arable land), 1967. (Source of data: June 4th Returns of the M.A.F.F., 1967.) (Map research by Mrs. A. Timmis) 55

FIG. 26. West Pembrokeshire: early potato production on selected farms: gross margin per acre. (Based on a survey made by D. Theophilus of the N.A.A.S. in 1967) 56

FIG. 27. West Pembrokeshire: first-lifting dates of *Home Guard* in 1967 57

FIG. 28. West Pembrokeshire: lifting dates and yields of the earliest field per farm. (The sample of farms used is identical with the one used for Fig. 27) 58

FIG. 29. Table showing the variations in weekly frosts and minimum temperatures at four sites in West Pembrokeshire, 1967 60

FIG. 30. West Pembrokeshire: irrigation required to July 1st, 1967 61

FIG. 31. Soil moisture deficits at Dale, Pembrokeshire, 1967. 62

FIG. 32. Lifting dates and yields from five Pembrokeshire farms growing *Home Guard*, June–July, 1967 64
FIG. 33. Table showing wind-throw damage to Forestry Commission plantations, 1962–8 68
FIG. 34a. Table showing fire losses in Forestry Commission plantations, 1954–67 68
FIG. 34b. Table showing fire losses in Forestry Commission plantations, 1962–7 69
FIG. 35. Proportion of years with timber losses of given extent due to fire (1956–67) 70
FIG. 36. Proportion of years with timber losses of given extent due to wind-throw (1962–8) 70
FIG. 37. Monthly distribution of timber losses due to wind-throw, (1962–8) 71
FIG. 38. Monthly distribution of timber losses due to fire: large fires, 1962–7; all fires, 1929–57 72
FIG. 39. Distribution of fire starts by time of day, 1929–57 (after Charters) 72
FIG. 40. Indicative distribution of wind by hourly wind speed for the year for a very windy site (after Golding and Stodhart) 73
FIG. 41. The relationship between critical wind speed and tree height for four principal soil types (after Fraser) 75
FIG. 42. Loss in present value (D.R.) when rotation is terminated at the indicated top height compared with optimum rotation 77
FIG. 43. Maximum value of investment in drainage, and expenditure on drainage, and from 5 ft 10 ft increases in height, in crops of different ages 77
FIG. 44. Present value of returns when the forest crop is grown to various ages, at 0, 10 and 20 years (Sitka spruce) 78
FIG. 45. Returns from thinned and unthinned crops. (Returns discounted at the beginning of the rotation) 78
FIG. 46. Comparison of community benefits with farm profits for various approaches to weather risk 110

Editor's Acknowledgements

SINCE 1958 the Editor has convened annually at Aberystwyth a symposium in aspects of agricultural meteorology. The eleventh in this series held in March 1968 adopted for discussion the subject of "Weather Economics". The present volume is based on the eight papers given at the symposium and on the discussions they stimulated.

The Editor wishes to acknowledge the generous assistance of all who have contributed to the success of the symposia series and in particular the preparation of the material for this volume. Thanks are due to the Director and officers of the Welsh Plant Breeding Station for their hospitality and use of the conference room. The secretarial assistance of the following is kindly acknowledged: Mrs. Sylvia B. Taylor, my wife, for abstracting the tapes, Mrs. Anne Timmis (who also compiled the indexes), Mr. Brinley Jones, Mrs. Mair Jenkins and Miss Pamela Chatwin. The cartography was done by Mr. Morlais Hughes and Mr. Michael G. Jones; the photography was done by Mr. David Griffiths.

The Editor is also most grateful to all the contributors to this volume for their effective and speedy co-operation. Mr. Dyfri Jones and Mr. John Aitchison contributed much in private discussion to the preparation of Chapter 9. A special thank-you is due to Mr. Neil Rubra who has not only advised throughout in detail on the preparation and editing of manuscripts but has also added an "Economic Postscript" (Chapter 10).

Finally, the Editor wishes to thank Mr. Alan J. Steel and his associates at the Pergamon Press for their unfailing courtesy and help at all times and for producing such an attractive volume.

Aberystwyth JAMES A. TAYLOR

Aberystwyth Memoranda

LIST OF PUBLICATIONS (October 1968)

Aberystwyth Memoranda in Agricultural Meteorology (edited by James A. Taylor)

Memorandum No. 1. *The Growing Season* (1958) (5/– plus postage).
Memorandum No. 2. *Shelter Problems in Relation to Crop and Animal Husbandry* (1959) (5/– plus postage). (Out of print.)
Memorandum No. 3. *Hill Climates and Land Usage, with Special Reference to the Highland Zone of Britain* (1960). (7/6 plus postage). (Out of print.)
Memorandum No. 4. *Aspects of Soil Climate* (1961) (7/6 plus postage).
Memorandum No. 5. *Climatic Factors and Diseases in Plants and Animals* (1962) (7/6 plus postage).
Memorandum No. 6. *Climatic Factors and Agricultural Productivity* (1963) (7/6 plus postage). Reprinted 1967 (12/6 plus postage).
Memorandum No. 7. *Major Weather Hazards affecting British Agriculture* (1963) (7/6 plus postage). (Out of print.)
Memorandum No. 8. *Climatic Change with Special Reference to Wales and its Agriculture* (1965) (10/– plus postage).
Memorandum No. 9. *Early Crop Production in the British Isles* (1966) (10/– plus postage). (Out of print.)
Memorandum No. 10. *Frost* (1967) (10/– plus postage). (Supplement containing discussion: 2/– plus postage). (Out of print.)
Memorandum No. 11. *Weather Economics* (1968) Published by Pergamon Press.

Orders should be sent *direct* to Mr. James A. Taylor (address below) to whom all cheques, etc., should be made payable:
James A. Taylor, Senior Lecturer in Geography, University College of Wales, Llandinam Building, Penglais, Aberystwyth, Cards., Wales, U.K.

Welsh Soils Discussion Group Publications

Report No. 1. *Forest Soils* (1960).
Report No. 2. *Glaciation related to Soils in Wales* (1961).
Report No. 3. *Soil Organic Matter* (1962).
Report No. 4. *Soil Moisture* (1963).
Report No. 5. *Land Unit Classification and Productivity* (1964).
Report No. 6. *Soil Fertility.*
Report No. 7. *Aspects of Soil Biology* (1965).
Report No. 8. *Upland Soils and Their Usage* (1966).
Report No. 9. *Trace Elements* (1967).
Report No. 10. *Methods of Soil Investigation* (1968).

N.B. These are *not* Departmental publications and must be obtained from: Hon. Secretary, Welsh Soils Discussion Group, c/o Welsh Plant Breeding Station, Plas Gogerddan, Aberystwyth, Cards., Wales, U.K.

Introduction

METEOROLOGY and economics make strange bed-fellows. Meteorology is a young and, confessedly, inexact science which has made marked advances in mathematical expression over the last 20 years, but has been slow to apply its established principles, for example, to studies of the weather-sensitive sections of the British economy. In British universities, meteorology is an underdeveloped subject represented in only a few institutions and is almost consistently and inevitably associated with the sciences of physics and applied mathematics and, very recently, environmental sciences. Economics, on the other hand, although being an even younger discipline, has expanded rapidly in nearly all British universities in the last 20 years or so and now forms a major component of the social sciences where it thrives in both its pure and applied forms. A veritable army of economists have been disseminated to responsible, decision-making posts in both the public and private sectors. This extra-version of the subject is in contrast to the confinement of meteorology to a few small niches in selected university faculties of science and to the quite restricted range of the Meteorological Office and its services.

In university departments of geography, however, it is traditional to study the perceptible relationships between the environment, its materials and processes on the one hand and social and economic activity on the other. Within this framework, analyses of meteorological data, related to both weather and climate, are placed within the full perspective of the biosphere. Economic systems are treated as locational phenomena within and dependent on, directly or indirectly, the biosphere which provides site, space and raw materials as required for potential economic exploitation. Whilst the primary controls of decision-making in these systems are clearly economic, technical, social and political, nevertheless the interpretation of the long-term validity and effects of such decisions must be referred as relevant to the full environment within which the systems operate. Thus emerges, for example, a sharp awareness of the impact of weather hazards on farm practice and farm economy, and of the need to measure such hazards in economic, as well as physical, terms.

The point has been well and truly made by some of the exceptional weather experienced during the British summer of 1968. The floods in south-west England in July were exceptional enough, but the September floods in parts of south-east England were even more disastrous. All the worst features of British floods and their consequences were starkly demonstrated. First, there was an inaccurate weather forecast. A trough of low pressure, which was expected to move southwards to the English Channel, in fact remained stationary, and the amounts and intensities of rainfall were much greater than anticipated. Thus, flood warnings were not issued promptly enough to warn potential victims, both those already in areas where flooding was imminent and also those travelling into such areas. For them and for the nation as a whole, the hydrological consequences of the indiscriminate planning and urbanization of the last 40 years—consequences predicted by a few, but ignored by most, authorities—became once again an inconvenient, painful and expensive reality. The concomitant underinvestment in catchment study and control during the same 40 years has, it is now realized, been criminal. It is encouraging, however, to note the current priority given to hydrological research by the Natural Environment Research Council, a priority which preceded, but which has been heavily underlined by, the recent severe flooding. This research will be of no avail until the prejudices of local authorities in resisting flood protection (or any other form of weather-proofing) on the grounds of immediate costs are properly exploded.

To quote some facts and figures, the estimate of uninsured losses (excluding public works) caused by the July floods in the West Country in 1968 is put at more than £2,000,000. The government of the day offered a mere £175,000 towards this, but voluntary subscriptions amounted to only £50,000. Previously, in the 1960 floods, claims had totalled more than £1,000,000 from the same area, of which the Exchequer found about four-fifths, but without prejudice to future claims. The principle emerges as to whether *ad hoc* disaster funds or grants should be replaced by a permanent national fund in the form of a social service.

The damage in the mid-September floods in south-east England has been estimated at £5,000,000. The insurance companies face claims totalling up to £12,000,000, which is more than twice what was paid out on the occasion of the 1953 floods on the east coast. The hurricane which crossed Scotland on January 15th, 1968, cost about £8,000,000. The only other more expensive weather hazard was the cold winter of 1962/3 which cost the insurance companies £20,000,000. Of the £12,000,000 of outstanding claims for the flood damage in south-east England this summer, some £5,000,000 will be paid out on householders' policies. Of the remaining £7,000,000, large sums will go to industrial and commercial groups and only a small sum to insured, individual farmers who rarely in fact take out this type of insurance policy. The overall costs to agriculture in drowned stock, lodged grain, rotting root crops, delayed harvests and damaged property are likely to reach six figures.

This is the scale of the cost of one occurrence of one type of weather hazard. So many other weather hazards register, often visibly but no less substantially, in farm accounts and in the accounts of industries which are more or less weather-sensitive, e.g. the construction industries, road haulage, etc. The total national cost if it could be worked out would be enormous: even regional costs have been shown to be alarming.

The papers and discussions which follow illustrate and measure the impact of weather hazards on the budgets of weather-sensitive industries such as farming and forestry. Reference is also made to other weather-sensitive industries and the effects of snowstorms on communications in built-up areas. Types of physical weather-proofing are discussed as well as the roles of financial insurance and subsidy against the risk of losses due to weather hazards. Attention is also given to the adjustment of day-to-day programmes, decision-making and available labour resources to particular weather probabilities assuming that the latter can be locally forecast with adequate accuracy and applied specifically to the industry or production process concerned.

As an economic variable, British weather has been relatively ignored and certainly underestimated in the past. This applies not only in practical agriculture but is equally relevant to the long-term research programmes in experimental plant and animal breeding, the end-products of which must, in the final assessment, be economic to the farmer and to the nation. Equally, it must be realized that many more aspects of industrial production are more or less weather-sensitive, be it directly via transportation, for example, or indirectly by the depletion of the labour force through certain illnesses affected by weather conditions.

The hope is expressed that the studies and commentaries in this volume will encourage more serious, long-term economic assessments to be made of weather hazards. Such assessments could be integrated, as required, with regional and national plans as applied not only to the rural areas of

Britain but also to the metropolitan areas since both, not least in hydrological and meteorological terms, are interrelated and interdependent.

Aberystwyth JAMES A. TAYLOR

CHAPTER 1

The Cost of British Weather

JAMES A. TAYLOR

BRITISH weather costs money. Numerous instances could be given. Despite the emancipation from environmental controls provided by modern technology and production techniques many systems within the economy remain, albeit selectively, weather-sensitive, especially if located in weather-sensitive areas. Indeed, it is the increasing effects of economic competition which have pushed productivities nearer to selected ecological margins, involving sometimes a greater vulnerability to weather variations. Duckham (1964) has illustrated this most clearly for farming systems; it must apply to all forms of economic activity associated directly or indirectly with stages of production at sites exposed to, or using materials or processes vulnerable to, weather factors.

The technique of cost/benefit analysis is bedevilled by the difficulty of measuring "imponderable" or latent factors, e.g. the amenity value to a community of a given site or area. In the sphere of weather economics the additional difficulty is added of the frequent lack of weather data immediately and precisely applicable to the economic activity or situation concerned. Despite the crudity and assumption involved in weather-costings, none the less several relevant examples can be quoted.

Firstly, for example, on the night of June 22nd, 1957, a late killing frost affected potato crops on mossland in Lancashire so severely that one farmer estimated an overnight loss of £1000, another of £2000. Secondly, in May 1950 three separate tornadoes developed along the Chiltern scarp overlooking the Vale of Aylesbury and did some £50,000 of damage to property, including crops (Smith, 1964). These are but two examples of relatively rare, extreme, isolated phenomena causing exceptional damage, for which it is usually regarded as uneconomic to cater either via weather-protection devices or by insurance against the weather. On the other hand, on the night of May 31st, 1962, a severe frost in the Upper Greensand country of Wiltshire was successfully countered by irrigation applied on the previous day and evening. It is estimated that the spraying of 38 of 45 acres of land saved a potato crop which was subsequently valued at £3000 (Hogg, 1966). However, relatively few farmers are able to afford or justify such relatively sophisticated protective techniques. Indeed, for certain extreme weather hazards virtually nothing can be done, particularly if there is little warning and more especially if the hazard is prolonged. The cold, long, snowy winter of 1947 is a case in point. In December 1946 the national sheep flock in Wales registered over 2 million (2,028,885) breeding ewes with ½ million (554,361) female lambs. By December 1947 the number of ewes was under 1⅔ million (1,626,574) and under ⅓ million lambs—decreases of 20 and 44 per cent respectively in numbers. The lamb crop in June 1947 was 940,000 which was about 750,000 less, i.e. over 55 per cent less, than normal expectation (Michael, 1962). Allowing £5 per lamb, this represents in itself a loss of £3,750,000 in one season. Again, more recently in January this year (1968) the extreme gales brought such havoc in central Scotland that the Forestry Commission estimate (Edlin, 1968) that trees representing 40 million hoppus feet of timber were blown down overnight. This is equivalent to 50 million true cubic feet of timber and is also about the same amount of timber yield as that from two average harvests in Scotland.† This is not a loss of timber; it means that large areas of trees were prematurely felled *en bloc* by the gales thus creating a premature harvest. At the very least the weather factor has intruded vastly in forest management and ultimately there are bound to be net losses in yield per stand from the affected areas which would otherwise have passed through successive stages of thinning. Contemporaneously, in the Glasgow conurbation total damage to property as a result of the hurricanes of January 15th, 1968, included damage to 70,000 corporation houses, which is half the total residences of that type in the area. Nineteen people were killed, 100 seriously injured and 1700 rendered homeless. This was an extreme event, but it indicates the extreme range which can be represented in British weather.

Again, with the "foot-and-mouth" outbreak in Britain apparently finally curbed at the time of writing (February 1968), one cannot resist the speculation (to be confirmed or otherwise by current research‡) that the

† Wardle (Chapter 7, p. 67) quotes a different estimate for this timber loss.

‡ P. B. Wright (*Weather*, June, 1969) pp. 204–13, concludes that "a large proportion of the spread" during the 1967/68 epidemic "was due to the wind".

effects of bouts of winds from the south and south-west just after the first outbreak of the disease at Oswestry on October 25th, 1967, may have been instrumental in its early spread northwards and north-eastwards. Inevitably, many more obvious and varied vectors were involved. Individual vectors are difficult to isolate. This is by far the worst outbreak since the disease become notifiable almost a century ago. It is likely that the number of animals lost is not far from the total number destroyed, because of the disease, in the previous 40 years. The costs involved are astronomical. The costs attributable to weather factors although probably minor could prove not to be negligible. Hurst (1968) has subsequently concluded from a study of outbreaks of "foot-and-mouth" disease in Britain over the last 30 years that "some outbreaks were associated with favourable meteorological conditions at a key time for transport from the Continent (e.g. 1952) while other distributions of the disease could not be so explained (e.g. Oswestry, 1967)."

On a more regular and annual basis the cost of clearing winter snows registers significantly in the budgets of local authorities. For instance, the county of Westmorland spends on the average over £50,000 a year keeping the Shap Fell route clear for traffic. £10,000 is spent on salt alone. Again, it has been estimated that up to 20 per cent and occasionally 25 per cent of the variations in milk yields from year to year in the counties of Wiltshire, Dorset and Somerset can be attributed to weather factors (Chambers, 1964). More comprehensively, the British Insurance Association pays out large sums each year, especially just after periods of prolonged snow and ice when road conditions cause delays and accidents. Between December 1962 and March 1963, which was the coldest winter for two centuries by some standards, the Association paid out more than £20 million in claims, many derived directly or indirectly from weather conditions.

The role of the meteorological services in the British economy has recently been summarized by the Director of the Meteorological Office (Mason, 1966). The economic value of the "civil national weather service" is put at £50–100 million per annum which for a cost of £4 million per annum gives a crude benefit/cost ratio of about 20 to 1. The estimated annual economic benefits to agriculture derived from the meteorological services are as shown in the table based on Mason on the next page. These are conservative estimates, particularly since comprehensive data for the whole country on agricultural economies achieved by taking weather advice are not available and would be extremely difficult to

Animal diseases	£4 million
Potato blight	£1 million
Sugar beet virus	£1 million
Cereals	£2 million
Hay and milk	£2 million
Total	£10 million

obtain. Mason, with reference to the building industry, estimates that 3·5 per cent of production per annum is lost to bad weather. The comparative figure for the United States, where more elaborate protective measures are taken, is 2 per cent. The U.K. figure may appear minor, but it represents a loss of £100 million on average per annum. The Central Electricity Generating Board also receives regular meteorological advice since the national and local demand for electricity is very sensitive to changes in temperature, especially rapid changes. For instance, a fall of 1°C (1·8°F) increases the load by 1·3 per cent in summer and 1·8 per cent in winter. Thus in fact an error of say 2°C (3·6°F) in a forecast could underestimate demand by as much as 1000 megawatts on a winter's day. The C.E.G.B. guards against this by maintaining a reserve capacity which costs about £2 million per annum. Without the benefit of meteorological advice, the Board estimates that the reserve capacity would have to be increased by at least 10 per cent. Again, since gas supplies have been used for central heating systems, that service too is sensitive to weather conditions. By using meteorological advice to adjust storage capacity to a practical minimum, an estimated £250,000 is saved nationally per annum. These are convincing illustrations of the impact of weather variables on industrial as well as agricultural output and efficiency.

In conclusion, the language of the economist provides a means of evaluating weather factors as related to selected producing systems. In this way the relationship between environmental conditions and economic usage becomes visible. However, such visibility is liable to be conditioned by variations in management and technology and not least by variations in the economic values themselves which are all essentially relative rather than absolute. These *caveats* aside, the economic measurement of weather hazards is not only possible and practical, but it enables comparisons to be made and greater efficiencies to be achieved. Moreover, it brings home the fundamental identity between economic and ecological studies, both of which are concerned to measure the turn-overs or budgets of selected systems. There remains, however, the need to reconcile short-term maximization as sought by the economist with long-term balance as advocated by the ecologist.

REFERENCES

CHAMBERS, R. (1964) Weather hazards and milk production: a review. *Memorandum No. 7 University College of Wales*, Aberystwyth, pp. 34–41.

DUCKHAM, A. N. (1964) Weather and farm management decisions. *Memorandum No. 7, University College of Wales*, Aberystwyth, pp. 10–24.

EDLIN, H. L. (1968) Private communication.

HOGG. W. H. (1966) Meteorological factors in early production. *Memorandum No. 9, University College of Wales*, Aberystwyth, pp. 16–29.

HURST, G. W. (1968) Foot-and-mouth disease. *The Veterinary Record*, vol. 82, No. 22, pp. 610–14.

MASON, B. J. (1966) The role of meteorology in the national economy. *Weather*, vol. XXI, No. 11, November, pp. 383–93.

MICHAEL, D. T. (1962) Climatic factors and disease in sheep. *Memorandum No. 7, University College of Wales*, Aberystwyth, pp. 32–36.

SMITH, L. P. (1964) Weather hazards in agriculture: a survey. *Memorandum No. 7, University College of Wales*, Aberystwyth, pp. 1–9.

CHAPTER 2

The Effect of the Weather on Farm Organization and Farm Management

J. M. STANSFIELD

THE function of a farmer or manager is to make the optimum use of his resources such as land, labour, capital, credit and know-how in order to achieve the aims of the farming unit. Usually this will be to maximize profits, one exception being a farm used for teaching and research such as the one for which the author is responsible.

Only in a situation involving some uncertainty has management a discrete function (Johnson, 1954). This statement is beginning to have meaning in certain rather specialized aspects of agriculture in Britain. For example, intensive poultry, pig, or beef units with "controlled environment" housing are being successfully operated by staff unsupervised for long periods but working to a precise set of instructions. Other farm enterprises demand constant management inputs; a major reason for this is that they are weather-sensitive. Duckham (in Taylor, 1967) divided weather-sensitive decisions into four categories:

(i) land use,
(ii) enterprise choice,
(iii) enterprise organization and intensity,
(iv) day-to-day operational decisions.

This chapter deals with (iii) and (iv) and discusses the economic effect of weather hazards on the practical running of a farming unit. It shows (using examples from the farming year) how profitability is related to output which in turn is dependent on the timeliness of husbandry operations.

Capital investment is mentioned together with the effects of weather on labour morale.

The results of Farm Management Surveys carried out by provincial agricultural economists commonly show a wide variation in economic performance within homogeneous geographical areas. To simplify the presentation of results, farms are divided into "average" or "premium", as in Fig. 1.

	"Average" farms £ per acre	"Premium" farms £ per acre
Costs (excluding bought food and seed)		
Fertilizers	4·0	4·5
Rent and rates	4·0	4·0
Machinery	7·5	6·0
Paid labour	5·5 } 7·5	5·5 } 7·0
Unpaid labour	2·0	1·5
Sundries	2·5	2·5
Total	25·5	24·0
Output		
Gross output	37·0	40·5
Net output	31·5	37·0
Net output from crops and grazing livestock	29·0	34·5
Profitability		
Management and investment income	6·0	13·0
Investment in machinery, tillages, livestock and deadstock	43·0	38·5
Return on above capital	14·0 per cent	34·0 per cent

FIG. 1. Input, output, and efficiency measures for predominantly arable farms (with more than 50 per cent tillage). *Farm Business Data*, May 1967, Department of Agricultural Economics, University of Reading.

The major factors contributing to higher profitability of the "premium" farms are the higher gross sales, marginally lower costs and lower investment. It is generally acknowledged that these higher outputs are achieved by better managerial ability. Management is often defined as a combination of observation, analysis, decision, action and acceptance of responsibility. The manager of a "premium" farm is usually good in all respects, especially action. Delays, particularly in decision-making, often mean that it is already too late. The baling of feeding straw directly behind the combine harvester before a shower of rain in order to maintain quality, particularly palatability, is a good example.

The theory and mathematics of decision-making are being widely studied (McQuigg, 1965). The role of the farmer as decision-maker is also being studied (Duckham in Taylor, 1967), but no generally accepted, straight-forward technique yet exists for measuring his management ability. Often the successful farmer makes good subjective decisions with regard to the weather if the alternatives are few and not too complicated, such as deciding when to cut a hay crop. In the future, farmers along with their advisers will be using more sophisticated techniques such as statistical simulation to solve their management problems. A cereal grower, contemplating the purchase of combine harvesters, grain drying and storage facilities has decisions to make involving more than one weather variable in addition to labour, machinery, credit and price factors. Some of the more complicated decisions are in Duckham's categories (ii) and (iii), and related to enterprise choice and organization. With this type of decision the farmer has time to study alternatives as well as the availability of a wealth of technical and management advice. A farmer about to buy a very large tractor costing £2000, for instance, could consult his National Agricultural Advisory Service machinery officer. Being a specialist, the adviser would already have worked with model situations and would know the minimum acreage needed to justify such a machine. On the farm, he would check such details as soil-type and find out the willingness of tractor drivers to work overtime before making a final recommendation to purchase.

OPERATIONAL DECISIONS

It is on decisions in category (iv), i.e. on a day-to-day basis, that the farmer must largely make up his own mind independently. He has limited time to analyse the situation, make a decision and implement it. Weather forecasts can certainly be of help and could perhaps be more widely used by individual farmers. The questions a farmer must ask himself with regard to category (iv) decisions are:

(a) Are the time and weather biologically suitable, e.g. for drilling?
(b) Are the time and weather physically suitable, e.g. for ploughing?
(c) If the answers to (a) and (b) are yes, what labour and machinery are available to do the operation?

If the farmer is a good decision-maker or if he has good weather advice available he ought to need less capital in his business or obtain a better return on the amount already invested. It is interesting to note that although the farming industry has in the National Agricultural Advisory Service a free and high-quality service, a number of farmers are prepared to pay fees in order to obtain the services of a private consultant. Many reasons have been given for this, a major one being that the farmer can call for assistance just when he requires it, often to take some responsibility with regard to many decisions in category (iv). There may well be a wide demand for extending such a service offered by private consultants. A telegram or telephone call could inform the individual farmer that the time had arrived to begin a particular, vital operation.

Local radio is widely used in the U.S.A. on a daily basis in helping growers with regard to the timing of their husbandry operations. As soon as local stations are established in this country, it would be of benefit to allocate the District Adviser a regular and frequent broadcasting session.

Throughout the farming year, many weather-dependent husbandry operations are critical with regard to timing and their effect on yields. Thus, for the autumn sowing of a wheat crop, Croxall and Smith (1965) have correlated yields for different planting dates with the fall in the 4 inch soil temperature for 0900 hours during the autumn. Their findings show that the variety *Capelle* should not be sown earlier than 10 days after the 4 inch soil temperature first falls below 55°F (12·8°C) and not later than 45 days after this point. They suggest that failure to plant within this period will result, on average, in a reduction in yield of 2 cwt per acre per week.

The majority of farmers with modern equipment can plough all the land required for spring planting before the end of December, especially if tractors are fitted with weather-proof cabs. Finney (1968) found that on the chalk soils of the Berkshire Downs, except for days when snow covered the ground, such equipment could plough continuously in one field or other on the farms until the New Year.

Spring seed-bed preparations usually commence as soon as soil conditions dry out sufficiently; drilling often follows within days. Charlton (1965), reporting on work at Experimental Husbandry Farms, quoted again the figure of 2 cwt per acre per week for loss of yield when drilling was delayed after mid-March. Dates of harvesting have also been shown to be critical in spring-sown barley crops at High Mowthorpe E.H.F. (1966). Timeliness is just as important with root crops as with cereals, particularly the planting and spraying dates. For example, Reading University farms have in recent years grown potatoes on contract for crisp manufacture, second earlies, the variety *Red Craig's Royal*, being grown for lifting in the second half of July. In order to ensure a crop of at least 8 tons per acre by that date several husbandry techniques are used to reduce the risk of weather hazards. Chitted seed is planted in early March on a light, sandy soil. Irrigation is used as frost protection as well as to correct moisture deficits. A rain canopy can be erected over the mechanical harvester to shelter the staff so that harvesting can continue uninterrupted by weather. These items involve capital expenditure but for second early potatoes a gross margin of £100 per acre justifies this outlay.

Many examples can be given of the effect of weather hazards on grassland production and utilization. On the whole more problems arise with conservation than with grazing management, the fall-off in digestibility of grass being of major importance.

The above examples emphasize the need to tackle the job just as soon as field conditions allow. Equipment must be of an adequate capacity to cope with the acreage and be in a good state of repair before the season's work commences. Off-peak periods should be used to service equipment, and a check should be made to ensure that seed, fertilizer and other items have been delivered before being required. A useful aid to the operator at planting time is a work list or chart. This can include details of field, acreage, variety, seed rate, total seed required and drill setting. Spaces are left for the operator to record date sown, seed used and actual drill setting.

Capital expenditure on livestock housing is often difficult to justify solely for protection from weather. The housing of sheep is a subject receiving much attention at this time. The two main types of farmer who need consider such an investment are first, the hill sheep farmer with the problems of wintering his ewe-hoggs, and second, the lowland farmer who in order to be sure of good pastures for later in the year takes his sheep off the land in January, February and March. On our farms at Reading, grassland output has certainly been increased by inwintering. The benefits have been improvements in the yield and quality of silage from leys which previously have been badly poached in the early months of the year.

The erection of a new dairy unit will commonly increase the annual fixed costs by a figure of £15 per cow, which at 3/– per gallon is 100 gallons of milk out of a lactation of 800 gallons. Weather protection alone certainly in south-east England could not directly be credited with such an increase in performance. It is often the indirect benefits, such as reduced poaching of pasture and reduced feed wastage, which help to justify the building costs. One very important justification for buildings is that they usually provide better working conditions for the staff. The industry is faced with a continually dwindling labour force, as staff leave the land not only for the higher incomes but also for the better working conditions found in other industries.

Weather hazards can seriously affect the morale of farm staff, which lowers their working efficiency. A prolonged sugar-beet harvesting season puts a strain not only on machinery and equipment but certainly also on men and management. The unpredictability of weather hazards is a real challenge to anyone managing a farm. If the holding is carrying a suitable balance of enterprises under categories (i) and (ii), decisions in the short-term, viz. categories (iii) and (iv), will be much easier to achieve. Capital can be seen to have a valuable role to play in reducing weather risks, as can the optimum timing of field work.

The economic effects on the individual farm will ultimately depend on the skill and ability of the manager or farmer. He will no doubt have more and better weather data in the future, together with adequate training in the techniques of day-to-day decision-making.

REFERENCES

CHARLTON, R. (1965) *Farmer's Weekly*, January 1st, p. 57.

CROXALL, H. E. and SMITH, L. P. (1965) Sowing dates for winter wheat. *N.A.A.S. Quarterly Review*, No. 68, pp. 147–9.

DUCKHAM, A. N. in Taylor, J. A. (Ed.) 1967) *Weather and Agricufture*, Pergamon Press, pp. 69–78.

FINNEY, J. B. (1968) Personal communication.

HIGH MOWTHORPE E.H.F. (1966) *7th Annual Report*, p. 32.

JOHNSON, G. (1954) *Managerial Concepts for Agriculturalists*, Lexington, Kentucky, University of Kentucky, p. 56.

MCQUIGG, J. D. (1965) Foreseeing the future. *Meteorological Monographs*, vol. 6, No. 28, pp. 181–8.

CHAPTER 3

Weather and Machinery Work-days

C. V. SMITH

FARM WORK AS A WEATHER-DEPENDENT OPERATION

Consider a question that must come up on most farms from time to time, namely "What size of tractor shall I buy"? The meteorological import of this is apparently nil. In other words it is nil until one begins to inquire what is expected of the tractor and to inquire, for example, at what rate will it be expected to work.

A rate of working implies a dependence on time and in the present context a suitable rate is specified by the ratio,

$$\frac{\text{peak work load (i.e. acres to be covered)}}{\text{probable number of days available to see the work through}}$$

Now timeliness is of essential significance in field operations. A delay at a given moment is likely to have repercussions all through the year—a late sowing means a late harvest, a late harvest means late autumn cultivations, and perhaps finally spring wheat instead of winter wheat will have to be sown. The farming calendar and the natural divisions of the farming year are recognized. The broad limits are known of the periods to which certain operations should be confined for them to be worth while. What has to be specified is the number of days within these periods that are likely to be available for work.

It is agreed that work on the farm is a weather-sensitive exercise. In the main the ability to push on with the field operations necessary between seed-time and harvest is more dependent upon current weather, and in particular upon recent and current rainfall, than perhaps on almost any other factor. Certain field operations may not be delayed by a small amount of rain and obviously the definition of a *dry-day*, i.e. one on which work is not impeded by rain, should properly vary with the project in hand and the time of the year.

WEATHER AND DECISION-MAKING

A little thought will show that perhaps the meteorologist should have some useful comment to offer at any of the stages where decisions on work at the farm are being made. Consider the planning stage, where the concern is the level of investment in manpower and machinery necessary to cope with the work load stipulated (perhaps over more than one season ahead). The weather data of past years will obviously be relevant, on the assumption that past weather statistics represent a population from which future years will show no significant deviation. In any one season certain specific operations will be necessary. A record of the recent and current weather can be a useful pointer to their timing. Examples such as planting date, irrigation requirements, the need for a spray programme, spring to mind.

Though it may be known which husbandry operations would now be expedient, the likelihood or the worthwhileness of carrying them out may turn on the weather that is expected or forecast.

A PRE-REQUISITE FOR THE WEATHER/WORK DECISIONS

There is one very basic assumption in the foregoing, namely that some correspondence may be detected between the weather variables and the field operations. It follows that the weather events after which the given crop should be treated or the weather period during which the given machine will, or should be allowed to, operate, have to be brought down to fairly precise meteorological terms. In other words, the critical or limiting values of the meteorological variables have to be specified for each operation. Obviously, to establish these threshold values initially, both weather and the field or biological data should be available simultaneously. Now it is true that whilst weather observations from a particular field on a particular farm may not be available, macro-scale weather observations will always be obtainable from the network of official weather stations. An estimate of the applicability of weather records at one site, to the weather likely to be experienced at another, is not an uncommon exercise.

What is almost certain to be lacking, at least with the degree of detail available in the meteorological observations, are field records of work done or of the state of the crop. This is the common situation and it appears that it falls to the meteorologist to bridge the gap and to define the weather that permits of, or suggests, a particular field operation. Purely empirical or statistical approaches to these problems are perhaps best avoided and the meteorologist is required to have some understanding of what the field operation sets out to achieve. It then becomes possible to reduce the processes involved in a successful operation to their physical essentials and to phenomena with which the physical processes of weather may be associated.

A COMMENT ON THE DATA

Though this approach may suggest which of the observed (or simply derived) meteorological variables are likely to be relevant, in the final choice of the weather elements employed in the correlation, some weight should be given to the way in which meteorological information is conventionally summarized and to the parameters which are readily available. If, for example, rainfall and soil moisture deficit appear equally valid parameters, it is far more convenient to work with rainfall data. Analyses and summaries are conventionally made of single meteorological parameters, e.g. rainfall alone, temperature alone. If concurrent requirements for more than one meteorological parameter are specified, e.g. minimum temperature and minimum relative humidity, then recourse must inevitably be made to the original, full meteorological observations. Summaries will have to be prepared individually, on an *ad hoc* basis. Even within the Meteorological Office, in this mechanical age, the tedium of counting is scarcely to be avoided.

The samples with which one has to deal in establishing the relation between weather and field work are likely to be small. Those relating to the field work are likely to be internally inconsistent in some degree and will perhaps be subjective in origin. Obviously, one can do no more than the data will allow. What is likely to be achieved is not precise relationships which will predict whether any given farmer will work on any given day, but simply a first approximation to this. Perhaps the best that can be hoped for is that the total number of work-days in a period of the order of 10 days should be indicated, with an error of not more than 1 or 2 days.

SOME SPECIFIC EXAMPLES

The discussion so far has been quite general and it is of interest to consider one or two specific requirements for long-term planning data that have been put by the N.A.A.S. (the National Agricultural Advisory Service) in the regions to which the author is attached. The first concerns a feasibility study for a grass-drying plant and illustrates the case where the conditions for successful field operations are known only within very wide limits. Broadly, the questions posed could be brought down to the number of cuts of grass which may be reduced by wilting in the field from an initial moisture content of the order 80 per cent (wet weight basis) to a moisture content of the order 68 per cent (whereby the output of the drier is increased by a factor of 2). The number of cuts each month was required, together with some estimate of the variation in the number of cuts over the years. The harvesting machine was not specified and crimping or crushing cannot be assumed.

Such was the problem. The only field data available are the general farming experience that cut grass may be brought to a moisture content of 40 to 60 per cent (wet weight basis) after a period of 8 to 48 hours in the swath in "dry" weather (tedding assumed). No supplementary information is available on the effects of field relative humidity, temperature, sunshine, or of small amounts of dew and rain.

What is known of the physical processes involved in the drying? Fortunately, the general form of the equilibrium curve between the relative humidity of the ambient air and the moisture content of grass is available. This shows that with air at 90 to 95 per cent R.H. the equilibrium moisture content of cut grass is 30 to 35 per cent (wet weight basis). From this it is inferred that when there is no free surface moisture, some initial drying of the grass will take place under most weather situations. If a mean period of 24 hours in the swath is adopted for the requisite drying and, assuming that grass will be cut only on a day that is dry and carried home only on a day that is dry, then the number of cuts brought in without complication becomes equal to the number of two-day spells of dry weather. This gives our first approximation to the information required. If a suitable sample of the weather of past years, for the area in question, is made available, the number of two-day spells in a given month may be presented in the form of a cumulative percentage frequency curve (F2) as in Fig. 2. The vertical axis has been relabelled to read not percentage frequency, but years in ten. In this way probabilities may be attached to the number of occasions when the grass may be cut and carried home. For example, in 30 per cent of the total of observations in May, or in 3 years in 10 in May, the number of cuts possible will be at least 10, or the probability that the number of cuts will exceed 10 is 7/10. In case farm advisers, from their own experience, were able to say that 24 hours in the swath in their particular areas was generally an over-estimate or under-estimate of the time necessary, then similar data on 1-day and 3-day dry spells was also supplied.

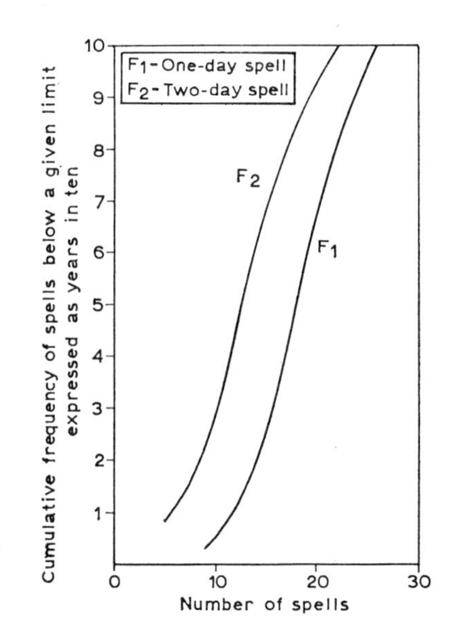

FIG. 2. Cumulative percentage frequency of dry spells in May: Nottingham (1916–50).

So far, a dry-day has not been defined in the present context. This was simply taken as one "without measurable rain". The first approximation is likely to be an under-estimate on two counts. Firstly, no allowance has been made for the timing of rainfall. In any 24-hour period, rainfall totals over the period 0900 to 0900 hours GMT are all that are readily available. It may be assumed that on average over all rain-days, the days are designated "wet" because of rainfall in the period 0900 to 2100 hours GMT, as often as because of rainfall in the period 2100 to 0900 hours GMT. If the grass is carried before dusk on the second day of a 2-day dry spell, rainfall after this time becomes unimportant. An upper limit to the number of such occasions is set by the number of 1-day dry spells, shown by the curve (F1) in Fig. 2. The second approximation to the information required is a curve set midway between (F1) and (F2).

Again, it is probable that small amounts of rain or dew may be tolerated on occasions and effective drying still achieved in the 2-day period. Against this it may be expected that not every period characterized by an absence of rain will be suitable for drying. The best that can be done without specific field observations is to marshal the factors that could lead to some bias in the final results. If these factors do not operate in the same direction it is hoped that some errors will cancel each other out.

A case will now be considered where the field data was known with a little more precision. The aim was to place gang work-days on some objective, quantitative basis. In the spring of 1966, the East Midland N.A.A.S. obtained returns of work done from about 35 farms on three broad soil types—*light*, *medium* and *heavy* land. Meteorological data were now required to be fitted to the farmer's decision of work no/work. What can be said about the physics of the problem? At the end of the winter it may be assumed that the soil will be at field capacity and that the land drains are running. Trafficability and soil workability for the spring cultivations will depend primarily on the soil moisture status. At this time of the year, evaporation and evapo-transpiration from the largely bare soil will be small but nevertheless positive, and will provide some drying of the top layers. What is known of the movement of water in the soil suggests that if the top layer has dried, renewal of the moisture in that layer will be mainly from above, by current rainfall, rather than from below. Rainfall would appear to be a convenient parameter with which to work. The type of criteria devised to fit the decision of work no/work are shown in Fig. 3. It was of course necessary to define "potential work-days", that is days on which no work was recorded but on which the weather was obviously not a limiting factor. This had to be done in such a way that the issue could not be said to be prejudged.

As has already been hinted, because of internal inconsistencies in the field data, the criteria are taken to give results that are only a first approximation. They correspond with the total of work-days accumulated over the spring cultivation period, but should not be employed to predict that any given day will be a work-day on a particular farm.

Criteria were first devised on the basis of field data for the spring of 1966. They were applied to a 10-year record of spring work from an independent source (with "light" soil) and appeared to stand up. When field data for 1967 became available, they were reassessed and adjusted to

1. *A dry-day*

In February to mid-April a "dry-day" is one with rainfall $\leqslant$ 0·07 in. On and after April 17th a "dry-day" is one with rainfall $\leqslant$0·09 in.

2. *The initial start to the season's work.*

(a) The day when work begins shall be dry.
The day when work begins shall be one on which the ground is not frozen (at a depth of 4 in.).

(b) When, on and after February 7th a total of 7 dry-days has accumulated (independently of the occurrence of intervening wet-days) work may begin after a sequence of 4 further, consecutive dry-days.

3. *The termination of work*

(a) In February to mid-April rainfall $\leqslant$0·14 in. in one day does not interfere with a run of work. Such days are counted as work-days.
On and after April 17th rainfall $\leqslant$0·19 in. in one day does not interfere with a run of work. Such days are counted as work-days.

(b) In February to mid-April a run of work is ended by a rainfall $\geqslant$0·15 in. in one day.
On and after April 17th a run of work is ended by a rainfall $\geqslant$0·20 in. in one day.
If the rainfall of the day which terminates a run of work is $\geqslant$0·20 in., that day is not counted a work-day.
If the rainfall of the day which terminates a run of work is $\leqslant$0·19 in., that day is counted a work-day.

4. *The recommencement of work*

(a) The day when work begins again after a wet interlude shall be dry.

(b) In February to mid-April a day's rainfall of $\leqslant$0·14 in. does not interefere with a run of work.
With a day's rainfall in the range 0·15 to 0·29 in. work may begin after 1 dry-day.
With a day's rainfall in the range 0·30 to 0·49 in. work may begin after 2 dry-days.
With a day's rainfall in the range $\geqslant$0·50 in. work may begin after 3 dry-days.
On and after April 17th a day's rainfall of $\leqslant$0·19 in. does not interfere with a run of work. In this period:
With a day's rainfall in the range 0·20 to 0·29 in. work may begin after 1 dry-day.
With a day's rainfall in the range $\geqslant$ 0.30 in. work may begin after 2 dry-days.

(c) The dry-days specified in section 4(b) before work can begin again need not be consecutive. If, after the termination of work, subsequent days are wet, then the interval during which no work is possible is, of course, extended. In such a wet interlude, each successive day is taken as the starting point for the recalculation of the number of dry-days required (according to the stipulations of section 4(b)) before work can begin again. Since such counts may in fact overlap, the day adopted for the recommencement of work is that furthest in time from the day when rainfall stopped work.

(d) If the wet interlude (when no work is possible according to the criteria above) extends over a period of 5 or more consecutive days, then the following requirements must be met before work can begin again:
In February to mid-April, 3 consecutive dry-days must occur before work can recommence.

FIG. 3. Work Criteria applied to spring cultivations on "medium" land in the east Midlands.

fit the data for both years. The results of applying the criteria to rainfall stations with long-period records in the east Midlands are shown in Fig. 4.

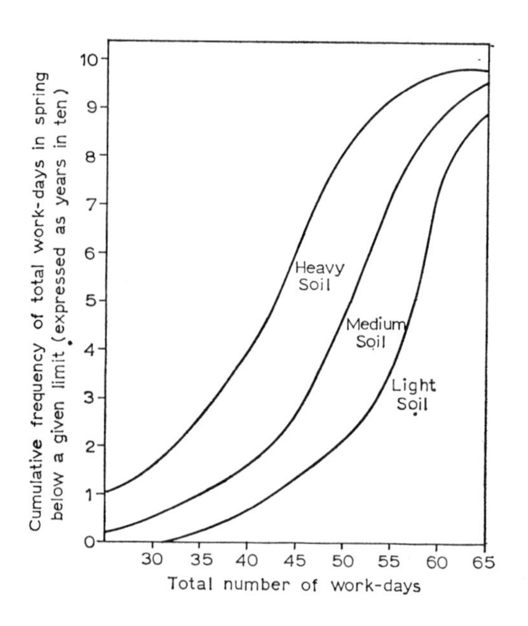

FIG. 4. Cumulative percentage frequencies of total work-days available for spring cultivations (February, March, April) for "heavy", "medium" and "light" soils in the east Midlands.

USE OF THE WEATHER/WORK CRITERIA

The immediate use of weather/work criteria is that they lead to long-term planning data. Curves of the form shown in Figs. 2 and 4 will indicate all possible or potential work periods. The user of such data has still to make some allowance for the efficiency of his work team. Aside from such questions as whether the team is available 7 days of the week, some estimate has to be made of the managerial skill of the farmer in seizing the opportunities for work that will be presented. How can we place a figure upon this? One way might be to assume that the farmer is guided in his decision to work or not by the current daily weather forecast and outlook. The efficiency of the farmer then becomes equated to the success in forecasting rain/no rain over 1-day or 2-day periods. An upper limit to efficiency for work requiring dry weather over 12 to 24 hours might then be no more than 80–90 per cent; for work requiring 2 days' dry weather, the maximum efficiency may be not more than 70 to 80 per cent. It is proposed to gloss over here the tactical use of current weather observations to decide that certain field work has become feasible or necessary. Some brief attention, however, should be given to the problem of relating projected field operations to the weather that is forecast. For, given that the labour force and the machinery available to it are adequate to cope with the vagaries of most seasons, the primary concern of the farmer must still be with what the current or coming season holds. It is surely reasonable to ask whether the criteria such as have been developed for spring work-days (and the long-period data derived from these) have any predictive value over the coming weeks. Is it possible to match such criteria to recent and expected weather and produce a forecast, for example, of the number of work-days likely to be available, without necessarily saying whether any given day will be a work-day?

Seasonal forecasts of weather are not yet available. Monthly forecasts are made, but statements on expected rainfall are very general. An indication of the number of rain-days is not attempted, though total rainfall amounts (within broad ranges) might perhaps be inferred. In any event, the available criteria are developed from daily rainfall amounts and a new approach would be necessary to work with monthly rainfall totals. Though at times a certain persistence can be demonstrated in the weather from month to month, no great reliance can be placed on the weather experienced in one month as a predictor for the next. In the same way, weather-dependent data such as work-days experienced in one month, will have no great value as a predictor of work-days in the following month. What of the applicability of weather forecasts over the period of a few days—at most a week? The weather up to the time of a broadcast is known and the current position of the weather/work relationship is therefore also known. If the weather of the coming week were presented in the form of the expected sequence of wet- and dry-days, or with some indication of the number of rain-days and perhaps some estimate of the rainfall total for the week, then from data such as these, estimates of the work-days available over the coming week might be possible.

This raises the question as to whether meteorological forecasts specifically prepared for the farming community should consist solely of the general weather prospects, as at present, or whether the agricultural implications should be specified. The argument that the individual farmer should make his own assessment does not hold universally. He does not have the necessary detailed and specialized information to do this (though the knowledge that he has of his own farm may on occasion modify any general conclusion to be drawn from agrometeorological forecasts).

ACKNOWLEDGEMENT

This paper is published with the permission of the Director-General of the Meteorological Office.

CHAPTER 4

Basic Frost, Irrigation and Degree-day Data for Planning Purposes[*]

W. H. Hogg

INTRODUCTION

The object of this paper is to provide some examples of the ways in which meteorological data can be adapted for agricultural or horticultural planning purposes. The examples chosen relate to frost, irrigation and degree-days data. The first of these is dependent mainly on local weather factors and is chiefly of importance in horticultural planning. The second depends on the more general factors of rainfall and transpiration, although local factors of soil type and the plant grown must also be considered; it has applications in both agriculture and horticulture. The third example deals with aspects of planning and management in heated glasshouses.

In this paper it is not intended to give full details of the use of meteorological data in planning; rather, a few simple examples are provided to indicate the possibilities.

FROST

The susceptibility of a site to frost, particularly in the spring, is a major factor influencing the location of horticultural holdings and of the fields selected for growing early potatoes on farms. Therefore the assessment of the suitability of sites forms an important part of the work of an agro-meteorologist in the southern half of England and Wales. It is, of course, possible to make qualitative assessments of suitability based on general meteorological principles. While these may help in making the choice between two possible sites, they provide no positive guidance on the economic consequences of growing on a particular site. For this, a numerical estimate of frost frequency is needed, which can usually be obtained only by a frost survey.

The object of a frost survey is to obtain probabilities of the occurrence of screen temperatures $\leqslant$32°F (0°C) after specified dates in spring; it is often valuable to use other base temperatures, e.g. 30°F (−1·1°C) or 28°F (−2·2°C), in addition to 32°F (0°C). The details of frost surveys will not be described here, but the following notes indicate the main steps involved.

1. Selection of sites: as far as practicable, this should cover the range from the most to the least frost susceptible on the land being surveyed. Usually some six to twelve sites are selected.
2. At each site a standard minimum thermometer is installed in an improvised screen (Gloyne and Smith, 1951) at about 4 ft height above ground.
3. Minimum temperatures are read each morning throughout the frost season, usually April–May.
4. The data are analysed on a weekly basis and for each week (starting week 1 on April 1st) the lowest, second lowest, and third lowest temperatures are selected.
5. These values (for one year only) are adjusted from the nearest long-term record to give estimates of long-term minima.
6. The estimates of long-term weekly minima are combined to give values from given dates until the end of the season. Thus emerges a value for the average of the weekly minima from April 1st to May 26th (8 weeks); also, corresponding values emerge for the second lowest and third lowest temperatures. The period April 8th to May 26th (7 weeks) provides similar data, and so on throughout the season.
7. The last stage is to convert these minimum temperature data into frost probabilities. This is done by reference to a long-period base station. Figure 5 shows curves for Long Ashton which relate weekly minimum temperatures to frost frequency (years in 10). The values given by using weekly minima and second and third lowest temperatures need smoothing, but the method gives rational, objective estimates.
8. In practice it is advisable to carry out the survey over two years, but the extension to three or more years makes little difference to the results.

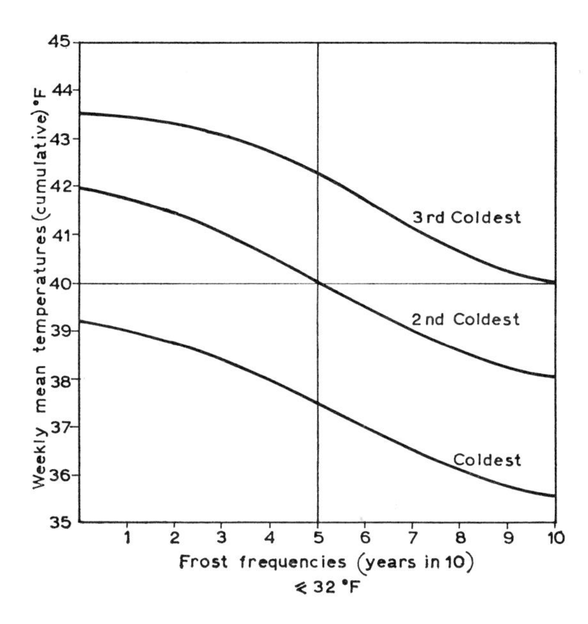

FIG. 5. The relationship between frost frequencies (years in 10) and weekly mean cumulative temperatures at Long Ashton, near Bristol.

Figure 6 gives the results of a survey carried out on a Somerset farm during 1963–5. This shows, for each of seven sites, the number of years in 10 with frost between the stated dates and the end of the frost season, here taken as May 26th. In order to give some idea of frost intensity, data are given for three temperature levels, viz. 32°F or below, 30°F or below and 28°F or below.

How can such information be used in planning? If it is assumed that for blackcurrants there will be a complete loss with a minimum temperature ≤28°F (−2·2°C) and a partial loss ≤30°F (−1·1°C) during April and May, then for land typified by site 4 there will be no crop in 5 years out of 10, half a crop in 2 years and a full crop in the other 3. If, apart from frost, a crop of 3 tons to the acre can be expected selling at, say, £110 per unpicked ton, the gross return is £330 per annum or £3300 over 10 years. Estimating to allow for reduced crops because of frost, the gross return over 10 years will be $3 \times 330 + \frac{1}{2}(2 \times 330) =$ £1320. It is somewhat unrealistic to ascribe all the reduction in yield to local site characteristics. The frost experienced in any area is the result of both regional and local factors, but unfortunately the resources available usually do not permit

Temperature	Apr. 1st	Apr. 8th	Apr. 15th	Apr. 22nd	Apr. 29th	May 6th	May 13th	May 20th
Site No. 1								
32°F (0·0°C)	10	10	9·3	8·5	7·8	7·2	5·4	4·4
30°F (−1·1°C)	6·2	5·7	5·5	5·2	4·7	3·7	2·0	1·4
28°F (−2·2°C)	4·2	3·7	3·2	2·7	2·3	1·8	1·1	0·7
Site No. 2								
32°F (0·0°C)	10	10	8·7	8·0	7·4	7·8	4·8	3·7
30°F (−1·1°C)	5·7	5·6	5·3	4·9	4·1	3·2	1·6	1·1
28°F (−2·2°C)	3·5	3·4	2·9	2·4	2·0	1·6	0·8	0·6
Site No. 3								
32°F (0·0°C)	10	10	10	10	8·8	8·2	6·3	5·1
30°F (−1·1°C)	6·4	6·0	5·8	5·6	5·3	5·1	2·5	1·7
28°F (−2·2°C)	4·4	4·0	3·8	3·5	2·9	2·5	1·4	1·0
Site No. 4								
32°F (0·0°C)	10	10	10	10	10	9·8	7·2	6·1
30°F (−1·1°C)	6·6	6·4	6·1	5·9	5·6	5·6	3·8	2·4
28°F (−2·2°C)	4·7	4·3	4·1	3·9	3·6	3·3	1·9	1·4
Site No. 5								
32°F (0·0°C)	10	10	10	10	10	9·8	7·5	6·3
30°F (−1·1°C)	6·3	6·3	6·1	6·0	5·8	5·6	4·7	2·6
28°F (−2·2°C)	4·3	4·3	4·1	4·0	3·7	3·4	2·2	1·4
Site No. 6								
32°F (0·0°C)	10	10	9·7	9·0	8·0	7·6	5·6	4·4
30°F (−1·1°C)	6·3	5·8	5·5	5·4	5·0	4·4	2·1	1·5
28°F (−2·2°C)	4·3	3·8	3·3	3·0	2·4	2·0	1·1	0·9
Site No. 7								
32°F (0·0°C)	10	10	10	10	9·5	8·5	6·5	5·3
30°F (−1·1°C)	6·4	6·0	5·8	5·7	5·5	5·2	2·7	1·8
28°F (−2·2°C)	4·3	4·0	3·7	3·6	3·3	2·7	1·6	1·0

FIG. 6. Results of a frost survey on a Somerset farm (1963–5.)

frost surveys over very wide areas to give regional results. The data from standard climatological stations may be reasonably used for this purpose and the values read from graphs similar to those in Fig. 5 are given in Fig. 7. Using these data for April and May the 10-year gross return is (5 × 330) + (2 × 165) = £1980 per acre. Even so, the attempt to grow on site 4

Temperature	Apr. 1st	Apr. 8th	Apr. 15th	Apr. 22nd	Apr. 29th	May 6th	May 13th	May 20th
32°F (0·0°C)	8·7	8·0	7·2	6·1	4·7	3·1	1·1	0·3
30°F (−1·1°C)	5·3	5·0	3·8	2·4	1·6	0·8	0·2	0·1
28°F (−2·2°C)	2·8	2·4	1·8	1·3	0·8	0·5	0·1	0·0

FIG. 7. Variation in regional frost risk, Somerset. Number of years in 10 with frost between stated dates and the end of the frost season.

land will diminish the annual gross return by some £66 per acre, or approximately the return from a half-ton crop.

A quite different question arises if the land is already planted with blackcurrants, namely whether it is economic to install frost prevention equipment to avoid at least some of the loss due to frost. A good frost prevention scheme should prevent all frost damage, overcoming both local and regional disadvantages. The estimated benefit per acre is therefore $\frac{£3300-£1320}{10}$, say, £200. Thus on 10 acres an average increase of £2000 would be expected and at, say, an estimated 20 per cent return on capital expenditure, an outlay of £10,000 would be justified. This sum might include the provision of storage, and the capacity of the reservoir would be determined by meteorological factors. For example, it is necessary to know on how many successive nights frosts can occur and, if possible, the length of time when air temperature is below freezing. These data cannot be obtained from frost surveys, but only from long-period data from the nearest climatological stations. For Somerset the data used are shown in Fig. 8 (Hogg, 1966).

Frost equipment, if installed, will certainly be used from early April and possibly even in March. A maximum of six successive nights can therefore be assumed, though if costs must be kept down, a design value of three nights would be reasonable. Data from Filton near Bristol (Hogg, 1968) suggest that it is unlikely that an average of more than 10 hours frost per night would occur during these periods. If, therefore, sprinkling is required for 10 acres and the proposed system will provide 3000 gallons per acre per hour, a maximum demand of $6 \times 10 \times 3000 \times 10 = 1{\cdot}8$ million gallons can be forecast, assuming no replenishment.

Date	Successive nights:					
	1	2	3	4	5	6 or more
April 1	6	5	2	1	1	1
April 8	5	4	1	1	1	1
April 15	5	3	1	1	1	1
April 22	4	2	1			
April 29	3	1	1			
May 6	3	1	1			
May 13	2					
May 20	1					

FIG. 8. Number of occasions with minimum screen temperature ⩽32°F (0·0°C) on one or more, two or more, etc., successive nights at Long Ashton, near Bristol, between the stated dates and end of frost season.

IRRIGATION-NEED

The experimental work of Penman (1948) established the principles of controlled irrigation in this country and practical methods have been developed in a number of Bulletins (M.A.F.F., 1954, 1962, 1967). These methods use water balance sheets, one of which is given for Oxford in Fig. 9. It refers to a grass surface or to some other "green cover" and to the growing season of April–September 1949. It is assumed that the aim is to keep the soil moisture deficit below one inch as far as possible. This is not always possible with a fortnightly balance, as there may be sufficient transpiration in a rainless fortnight to increase the S.M.D. well above 1 in. For example, during the fortnight August 16th–31st the deficit rose to 2 in. From this it is apparent that during 1949 some 10·5 in. of water were needed to ensure that the soil moisture deficit was kept (mainly) below 1 in. It can readily be seen that a farmer can keep his irrigation in step with the water needs of his plants provided he knows the rainfall, which he can easily measure, and has an estimate of water consumption, which can be provided for him.

However, our primary concern is the strategic, rather than the tactical, use of water balance sheets. In this simplified water balance sheet only two meteorological parameters appear, rainfall and sunshine. It therefore follows that for all places with long rainfall and sunshine records it is

Period		Total need pre-estimate	Actual rainfall	Rainfall deficit	Estimated soil moisture deficit (cumulative)	Planned soil moisture deficit	Irrigation needed	Estimated soil moisture deficit after irrigation	Sunshine correction	Corrected soil moisture deficit
April	1–15	0·95	0·76	0·19	0·19	1 in.	—	0·19	+0·11	0·30
	16–30	0·95	0·55	0·40	0·70	1 in.	—	0·70	+0·12	0·82
May	1–15	1·50	0·48	1·02	1·84	1 in.	1·84	Nil	+0·14	0·14
	16–31	1·55	1·93	—0·38	Nil	1 in.	—	Nil	+0·15	0·15
June	1–15	1·80	0·46	1·34	1·49	1 in.	1·49	Nil	+0·11	0·11
	16–30	1·80	0·00	1·80	1·91	1 in.	1·91	Nil	+0·12	0·12
July	1–15	1·85	0·25	1·60	1·72	1 in.	1·72	Nil	+0·34	0·34
	16–31	1·85	0·66	1·19	1·53	1 in.	1·53	Nil	+0·33	0·33
Aug.	1–15	1·55	0·99	0·56	0·89	1 in.	—	0·89	+0·10	0·99
	16–31	1·50	0·49	1·01	2·00	1 in.	2·00	Nil	+0·10	0·10
Sept.	1–15	0·80	0·12	0·68	0·78	1 in.	—	0·78	+0·03	0·81
	16–30	0·80	2·17	—1·37	Nil	1 in.	—	Nil	+0·03	0·03
Total							10·49			

FIG. 9. Irrigation-needs at Oxford, 1949.

possible to construct water balance sheets year by year and thus to obtain estimates of long-term irrigation-needs. For example, from the water balance sheets for Oxford over the period 1930–49 it is found that the total irrigation-need in 20 years is 126 in. In the driest year of the 20, 10·5 in, was needed, and some irrigation was needed in 19 of the 20 years. Using a computer it has been possible to construct water balance sheets for seventy-six stations over a 20-year period and to extract relevant data for mapping.

The water balance sheet given for Oxford (Fig. 9) may be taken as applicable to grass, a crop for which irrigation may be desirable at any time during the growing season April–September. For many other crops interest may be only in a restricted part of the growing season and, with the growing demand for water, it may be necessary to restrict irrigation to specific response periods for some crops. Also, because of variations in both rooting depth and in water-holding capacity of various soils, irrigation data may be needed for various irrigation plans or schedules.

Data have therefore been computed for every possible combination of successive periods of 2, 3, 4, 5 and 6 months, as follows:

Periods used for maps of irrigation-need.

April–May, May–June, June–July, July–August, August–September
April–June, May–July, June–Augist, July–September
April–July, May–August, June–September
April–August, May–September
April–Deptember

The four irrigation schedules used are given below:

Irrigation schedules used for maps of irrigation-need.

S.M.D.1. Allow S.M.D. to reach 1 in., then restore to zero (field capacity)
S.M.D.2. Allow S.M.D. to reach 2 in., then restore to zero (field capacity)
S.M.D.3. Allow S.M.D. to reach 3 in., then restore to 1 in. deficit
S.M.D.5. Allow S.M.D. to reach 5 in., then restore to 2 in. deficit

As seventy-six stations were used for a 20-year period with fifteen response periods and four irrigation schedules, the total number of balance sheets computed was 91,200. For each station the following information was selected from 20-year groups of water balance sheets (a 20-year

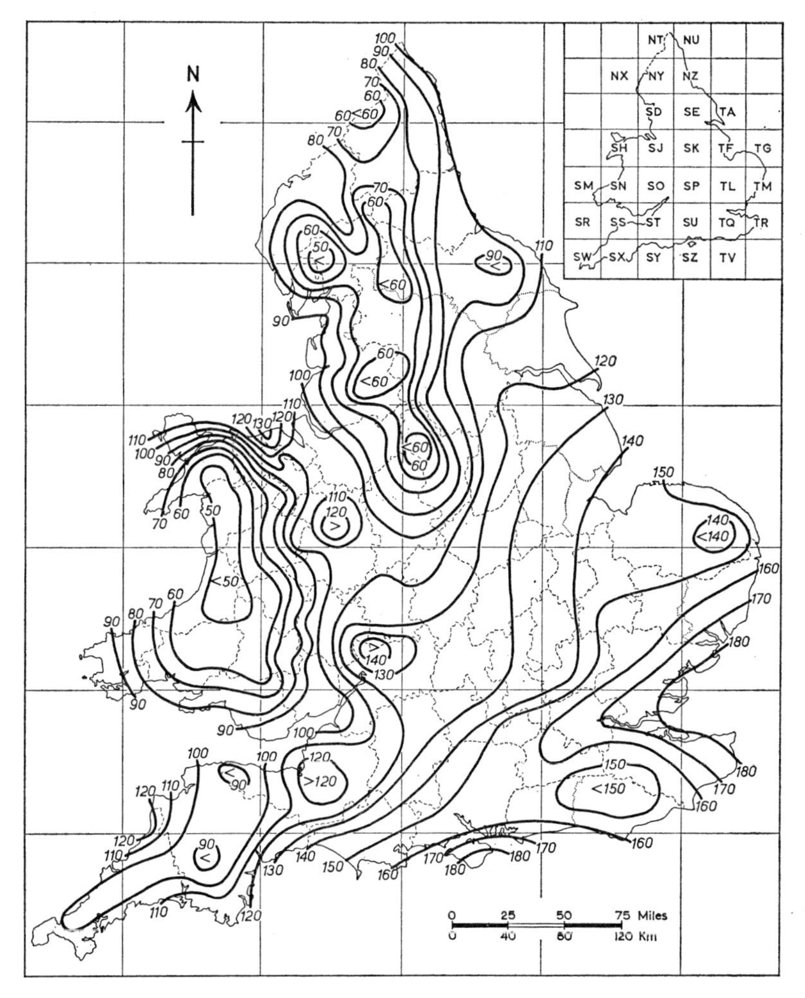

FIG. 10. Total irrigation-need over 20 years (in.). Preiod: April–September. S.M.D. 1 in.

group comprises the water balance sheets for 1930–49 for any given response period and irrigation schedule).

1. Total need in 20 years.
2. Adjusted total need in 20 years, i.e. the total need if the year with the fifth greatest irrigation-need is taken as the year of maximum need (the sum of the fifteen lowest values plus 5 × 16th value).
3. Need in driest year in 20, i.e. the year with the greatest irrigation-need in 20 years.

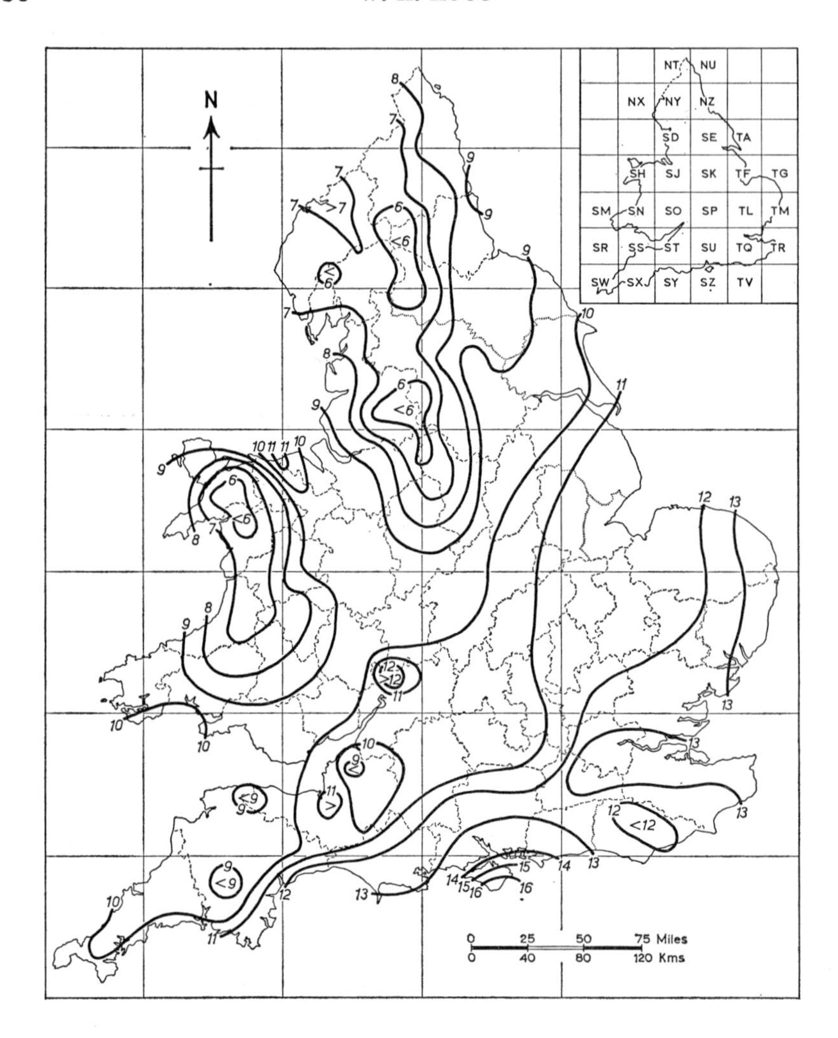

FIG. 11. Irrigation-need for driest year in 20 (in.) Period: April–September. S.M.D. 1 in.

4. Need in fifth driest year in 20.
5. Number of years in 20 with any irrigation-need.

The values were plotted on 300 base maps and isopleths were drawn. These have recently been published (Hogg, 1967) and can be used in long-term estimates of the water needed for irrigation. Modified versions of four of these maps are reproduced as Figs. 10, 11, 12 and 13.

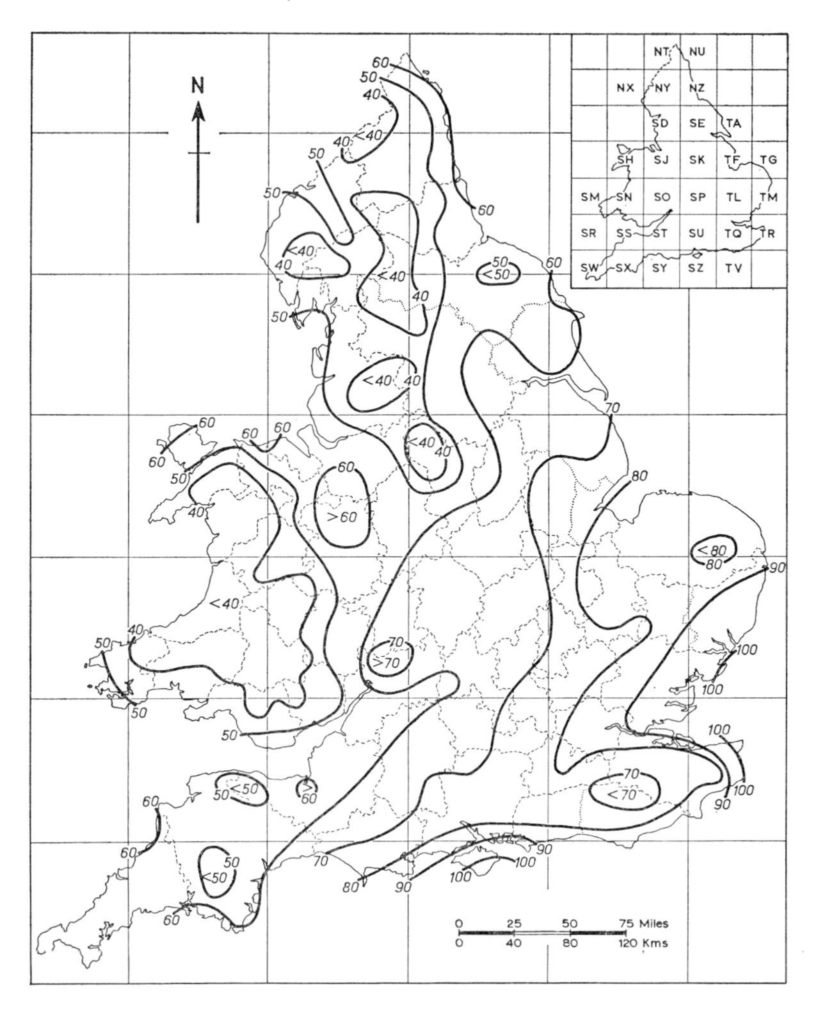

FIG. 12. Total irrigation-need over 20 years (in.). Period: April–June. S.M.D. 1 in.

In using the maps, decisions must be reached concerning which response periods and irrigation schedules are appropriate. These are husbandry decisions and need not be further considered here. The maps for April–September, S.M.D. 1 in. would be suitable for grass on medium or light soils and give the estimates for the Shrewsbury district shown in Fig. 14. Thus if it is proposed to irrigate 50 acres of grass the average annual need would be 300 acre inches and assuming an increased gross margin† of

† Gross margin is defined as the gross output of an enterprise less its variable costs.

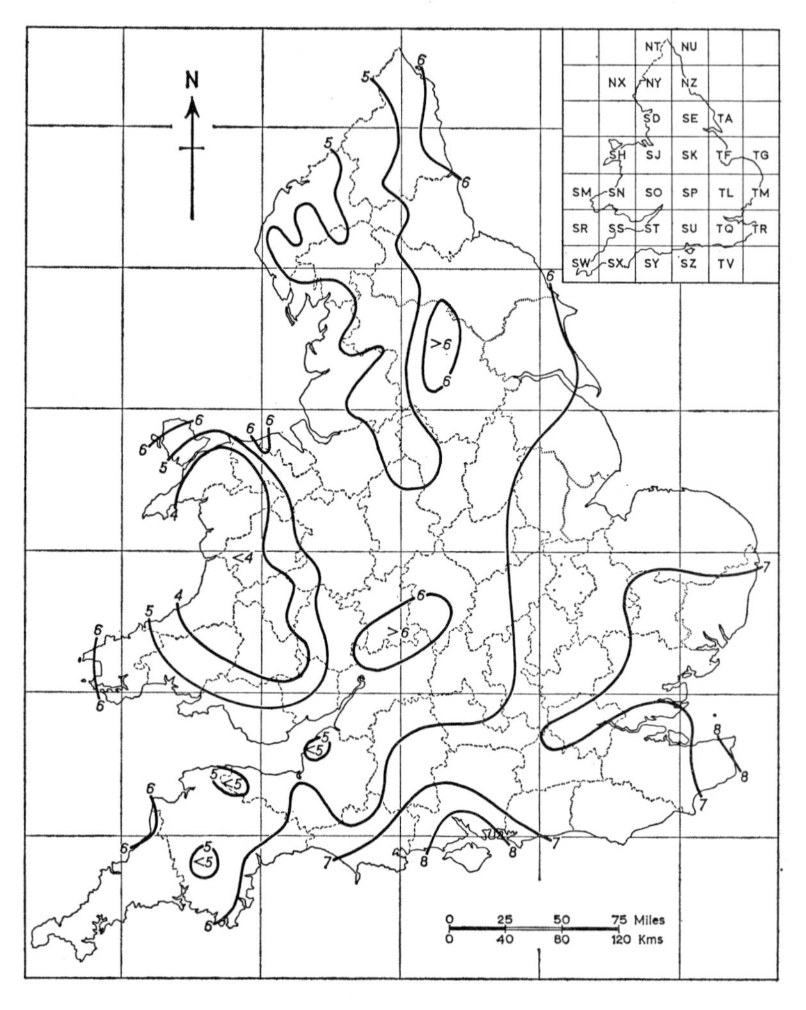

FIG. 13. Irrigation-need for driest year in 20 (in.). Period: April–June. S.M.D. 1 in.

about £5 per acre inch, the total increase could be £1500 which at, say, 20 per cent return would permit an expenditure of £7500. If the capital works were to include the construction of a reservoir it would have to provide a maximum of 450 acre inches, although its size would depend on many other factors such as method and rate of replenishment, evaporation and seepage. For some crops it may be better to use a design value based on the irrigation need in the fifth driest year and this would give 350 acre inches in the present example.

1. Total need (see Fig. 10)	120 in. in 20 years
2. Adjusted total need	110 in. in 20 years
3. Need in the driest year (see Fig. 11)	9 in.
4. Need in the fifth driest year	7 in.
5. No. of years with irrigation-need	19 years in 20

FIG. 14. Irrigation-needs, Shrewsbury, April–Sept., S.M.D. 1 in.

The figures in Fig. 15 have been taken from the maps for April–June, S.M.D. 1 in., and could be considered appropriate to early potatoes, a high value crop. Approximate values for the Pembrokeshire coast are:

1. Total need (see Fig. 12)	50 in. in 20 years
2. Adjusted total need	40 in. in 20 years
3. Need in the driest year (see Fig. 13)	6 in.
4. Need in the fifth driest year	4 in.
5. Number of years with irrigation-need	16 years in 20

FIG. 15. Irrigation-needs, Pembrokeshire coast, April–June, S.M.D. 1 in.

Using similar reasoning and assuming 5 acres for cropping and an increased gross margin of £30 per acre inch leads to a total assumed increase of £375 due to irrigation, justifying a capital expense of £1875 based on a 20 per cent return. The design value for a reservoir, based on the driest year, is 30 acre inches.

DEGREE-DAYS AND GLASSHOUSE HEATING

The use of degree-days is often a useful approach in dealing with heating problems. The concept of degree-days can best be understood by considering the temperature on a given day. If, on that day, the temperature is above a datum of $X°$ for n hours and the mean temperature during that period exceeds $X°$ by $m°$, there have been nm degree-hours above $X°$, or $nm/24$ degree-days above $X°$.

In practice, daily values of degree-days are derived from empirical formulae using daily maximum and minimum temperatures. A method has been devised for making estimates of the average number of degree-days above or below any stated base temperatures (Shellard, 1959) and more recently this method has been used to provide estimates for all glasshouse areas in southwest England and the west Midlands (Davies, 1966, 1967, 1968). These estimates give the average number of degree-days below bases at 5°F (2·8°C) intervals from 40°F to 70°F (4·4° to 21·1°C) on a half-monthly basis. An example of the type of information given by these estimates is shown in Fig. 16. These refer to Penzance, Cornwall, but are given for monthly intervals.

Month	Base temperature (°F)						
	40°	45°	50°	55°	60°	65°	70°
January	20	80	180	315	465	620	775
February	30	90	185	305	440	580	725
March	0	35	120	250	405	560	715
April	0	5	55	160	305	455	605
May	0	0	5	60	185	345	495
June	0	0	0	10	70	190	335
July	0	0	0	0	40	135	265
August	0	0	0	5	45	120	255
September	0	0	0	15	70	180	325
October	0	0	5	65	185	330	490
November	0	15	65	170	315	465	615
December	5	50	140	270	420	575	730

FIG. 16. Averages of monthly degree-days below stated bases at Penzance, Cornwall.

If, therefore, it is known at what temperatures crops will be grown and over what periods, the total heat loss from the structure can be calculated, which is the heat required to keep the air within the house at the required temperatures. The formula normally used is:

$$X = (EGA \times 1 \cdot 1) \times (D \times 24)$$

where X is the heat loss from the structure in BTUs, EGA is the equivalent glass area of the house, and D is the average number of degree-days below the required temperature(s) for the period considered. Such a formula can readily be used to estimate fuel consumption, knowing the calorific values of the various fuels (BTUs per lb or gallon) and the efficiency of the heating system. It can also be used as a design factor for the heating system, mainly to determine the size of the boiler. Since the boiler must be able to deal with extreme conditions of heat loss, e.g. very windy weather, the factor 1·1 is replaced by 1·4 for this purpose (M.A.F.F., 1964).

Degree-day data can also be used to apportion the fuel costs applicable to any given enterprise on a glasshouse holding, and thus to estimate the profitability of individual enterprises. A simplified example is given below, but it is first necessary to deal with a possible complication. The tables of degree-days referred to above have been computed on the assumption that the base temperature remains constant over a period of weeks. In modern commercial glasshouses it is now the practice to work to a lower temperature during the night than during the day and this clearly invalidates the direct use of tables. Also, the length of "day" and "night" in this context is variable, the night being considerably longer than the day in winter.

An examination of average hourly temperature curves for each month at stations in south-west England and south Wales has shown that it is possible to make good approximations by using standard data and weighting it according to the length of the day. For day lengths of 8, 12 and 16 hours it was found satisfactory to weight in the following manner:

Day-length	8 hours	12 hours	16 hours
Weighting	$\frac{D}{3} + \frac{2N}{3}$	$\frac{D}{2} + \frac{N}{2}$	$\frac{D}{2} + \frac{N}{2}$

where D is the number of accumulated degree-days on the assumption that the day temperature is maintained for 24 hours and, similarly, N the number worked for the night temperature.

A feasible example could be as follows. In a glasshouse in Gloucestershire tomatoes were heated from early February to mid-March, followed by cucumbers until mid-September. The temperature regimes are given for a 12-hour day. The computation of accumulated degree-days is given in Fig. 17.

Assuming a fuel bill of £100, some £32 could be attributed to tomato production and £68 to cucumbers. If more than one house is involved, it may be necessary to allow for differences of size and design, and also of the areas over which the individual crops are grown, but the principle of the method remains.

Other applications are possible. For example, a knowledge of the total heat requirements for each crop will enable theoretical fuel costs to be

Dates	Crop	Night temp.		N	Day temp.		D	$\frac{D}{2}+\frac{N}{2}$	Crop totals
		°F	°C		°F	°C			
Feb. 1–14	Early tomatoes	55	12·8	230	64	17·8	350	290	
Feb. 15–28		56	13·3	230	68	20·0	390	310	
March 1–15		59	15·0	250	68	20·0	385	320	920
March 16–31	Cucumbers	60	15·6	245	70	21·1	405	325	
April 1–May 15		62	16·7	580	70	21·1	925	750	
May 16–Sept. 15		65	18·3	635	70	21·1	1195	915	1990

FIG. 17. Example of computation of degree-days for apportionment of fuel costs.
Note: Degree-days are given only for the Fahrenheit scale.

computed. If these differ greatly from the actual costs it may be due to faulty management or to unsatisfactory houses, e.g. they may permit too much heat loss. Alternatively, of course, the difference may be explicable in terms of the weather of that particular season.

ACKNOWLEDGEMENT

This paper is published by permission of the Director-General, Meteorological Office.

REFERENCES

DAVIES, J. W. (1966) Accumulated degree-days below selected bases in the South-west Region. Meteorological Office.

DAVIES, J. W. (1967) Accumulated degree-days below 65°F and 70°F in the South-west Region. Meteorological Office.

DAVIES, J. W. (1968) Accumulated degree-days below selected bases in the West-Midland Region. Meteorological Office.

GLOYNE, R. W. and SMITH, L. P. (1951) Shielded thermometer mounts. *Met. Mag.*, vol. 80, pp. 203–4.

HOGG, W. H. (1966) Air frost in spring at Long Ashton. *Report Long Ashton Res. Sta. for 1965*, pp. 290–8.

HOGG, W. H. (1967) *Atlas of Long-term Irrigation Needs for England and Wales.* Ministry of Agriculture, Fisheries & Food.

HOGG, W. H. (1968) The duration of spring frosts on successive nights. Meteorological Office.

MINISTRY OF AGRICULTURE, FISHERIES & FOOD (1954) *The Calculation of Irrigation Need.* Technical Bulletin No. 4.

MINISTRY OF AGRICULTURE, FISHERIES & FOOD (1962) *Irrigation.* Bulletin No. 138.

MINISTRY OF AGRICULTURE, FISHERIES & FOOD (1964) *Commercial Glasshouses.* Bulletin No. 115.

MINISTRY OF AGRICULTURE, FISHERIES & FOOD (1967) *Potential Transpiration.* Technical Bulletin No. 16.

PENMAN, H. L. (1948) Natural evaporation from open water, bare soil and grass. *Proc. Roy. Soc. A*, vol. 193, pp. 120–45.

SHELLARD, H. C. (1959) Averages of accumulated temperature and standard deviation of monthly mean temperature over Britain, 1921–50. Meteorological Office Prof. Note No. 125.

CHAPTER 5

Variations in the Marginal Value of Agricultural Labour Due to Weather Factors

W. J. TAGGART

THE intention in this paper is to describe an attempt to isolate the effect that variations in the weather might have on farm profit, by affecting the availability of labour. It should be borne in mind that the implications of the results obtained will depend almost wholly on the accuracy with which the loss of time due to bad weather can be calculated. The number of hours suitable for outside work on an arable farm were calculated for three "weather situations", using the records of three rainfall recording stations near Edinburgh, and applying the criteria suggested by Taggart (1967). These three sets of workable time were then used successively to arrive, by linear programming, at three optimal plans for a single farm, the labour availability being the only variable parameter.

THE FARM

The farm used for the comparison is one of 403 acres, made up of 183 acres of medium land (*Dreghorn* soil series) and 220 acres of light land (*Frazerburgh* and *Macmerry* soil series). The possible crops, and their yields, are given in Fig. 18. A full-time staff of five men is employed, plus casual labour required for potato dressing in winter. Several livestock enterprises are possible, but no stockmen are employed, the work on stock being done by the general staff. The livestock enterprises considered are: winter fattened cattle; summer fattened cattle; 18 month beef; breeding sheep with lambs fattened; pigs fattened in cattle courts in summer.

Crops	Medium land yield	Light land yield
	tons/acre	tons/acre
First early potatoes	7	7
Second early potatoes	9	9
Early maincrop potatoes	12	10
Maincrop potatoes	12	10
Sugar beet	14	12
Wheat	2	1·6
Barley	1·85	1·5
Grazing		
Silage (1 cut)	8·4	7·5
Hay	2·8	2·5

FIG. 18. Crops considered for inclusion in the farm plans.

The linear programming model used includes: capital supply at 9 per cent interest; limitations imposed by a disease control rotation and by available crop storage and livestock accommodation; least-cost feed selection for livestock; various crop disposal alternatives; optional use of overtime and casual labour. Machinery selection was not included, in order to save computer time.

THE RAINFALL STATIONS

The records of three stations were used to provide the variations in weather. It should be noted that all three are within a relatively small area and that the pattern of farming does not vary markedly, between actual farms in the three districts. Station I is at 605 ft above sea-level, 6 miles south of Edinburgh. Station III is about 50 ft above sea-level, on the coast 23 miles north-east of Station I. A ridge of high ground rises to about 350 ft between the sea and Station II, which is 150 ft above sea-level, 7 miles south-south-west of Station III, and 16 miles east of Edinburgh.

For each 14-day period through the year, the number of days reckoned to be suitable for outside work on an arable farm is expressed as a percentage, using data from the three stations. These weather-dependent percentages (W.D. %) are compared in Fig. 19. The total time in any period, less the weather-dependent time, is assumed to be "wet", i.e. unsuitable for most outdoor work on an arable farm. It will be seen that the main differences in the weather-dependent percentage are in December, January, February and March.

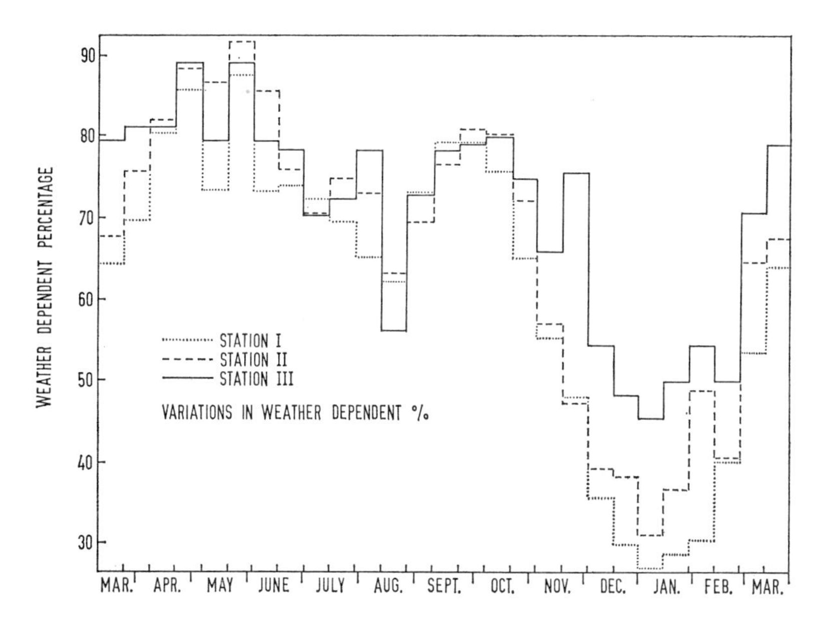

FIG. 19. Comparison of proportions of time suitable for outdoor work (periods of two weeks).

From this point, for convenience, the names Station I, Station II and Station III will be used to denote the three "farms".

In the model, the year has been divided into five seasons: spring, February 15th–April 15th; early summer, April 16th–June 20th; summer, June 21st–August 15th; harvest, Augist 16th–November 18th; winter, November 19th–February 14th. The average weather-dependent percentages for these seasons are shown, and inter-station variations compared, in Fig. 20 which possibly shows more clearly than Fig. 19 that the variations are mainly in winter and spring. In spring, Station I has 83 per cent of the workable time available at Station II, and in winter 61 per cent. During the harvest seasons the three Stations are very similar, Station I having 96 per cent of the workable time available at Station III.

As stated earlier, the farm staff is five men, and so limits are set on the total number of hours available in each season. The time available in each season is subdivided into "weather-dependent" and "wet" according to the appropriate W.D. %. Weather-dependent time can be used for "wet" weather work, if necessary (but not vice-versa), and overtime labour is available at the appropriate cost, if required.

The cropping in the optimal solutions for the three stations is compared in Fig. 21. It should be remembered that the variations in cropping balance are due solely to changes in the balance of labour availability, and not to any alteration in labour requirement or crop yield, although these will also be affected by the weather.

The profit from the farms varied in response to the effect of the three local climates on labour availability by increasing by £492 or £1·22 per acre from Station I to Station II, and by £810 or £2·01 per acre from Station II to Station III.

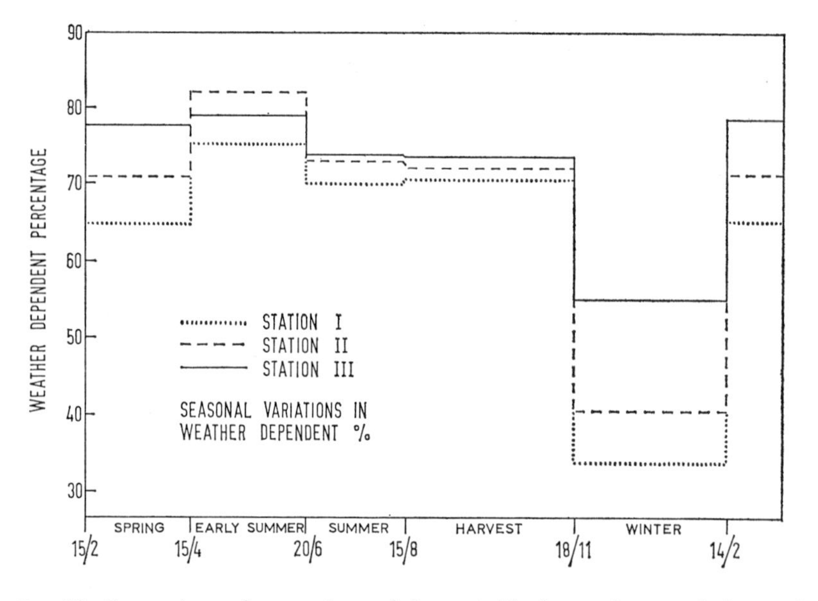

Fig. 20. Comparison of proportions of time suitable for outdoor work (seasonal periods).

One of the most useful attributes of linear programming as a planning tool is the amount of peripheral information which is available. In this case the value of additional weather-dependent labour is of interest. The scarcity value of one man-hour varies with the season, being nil in the early summer period, from April 16th to June 20th when the staff is underemployed, £0·98 to £1·02 in the summer period, June 21st to August 15th, when early potato lifting is under way, and £1·95 to £2·00 in the winter period. Now these indications are useful, showing the break-even value of labour, per hour, for weather-dependent work in these seasons, but in this form could relate only to the hiring of casual labour.

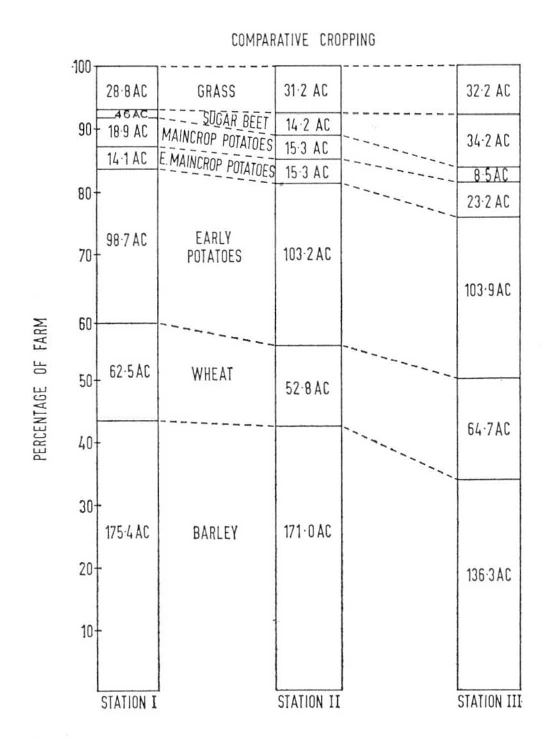

Fig. 21. Effect of labour availability variations on crop distribution selected.

However, if a full-time man is to be hired, then he has to be paid when his labour is worth nothing, as well as when his labour is worth £2 per hour. Therefore the annual break-even value of an additional man is calculated as demonstrated by Tyler (1966). For Station I, this is £1288; for Station II, £1355; and for Station III, £1516.

CONCLUSION

This paper is not intended to define the economic effect of the weather on farm profitability via labour availability, since it is not possible to generalize on this topic. However, subject to the assumption that it is possible to assess the effect of weather on labour availability, a possible method of measuring the economic effect of weather variations is suggested. The amount of profit variation found would seem significant, even in response to climate differences which, to a non-meteorologist, seem slight.

The results of the comparison are summarized in Fig. 22.

Station	Station I	Station II	Station III
Weather	Poorest	Intermediate	Best
% roots	33·8	36·7	42·1
% cereals	59·1	55·5	49·9
% grass	7·1	7·8	8·0
% of farm Gross Margin coming from livestock	19·2	19·4	19·7
Profit per acre increase over Station I	£0	£1·22	£3·23
Break-even annual value of an an extra man	£1288	£1355	£1516

FIG. 22. Comparative summary of enterprise distribution, profitability, and the marginal value of labour.

REFERENCES

TAGGART, W. J. (1967) Farm work planning and the adjustment of labour availability to allow for weather loss. *Farm Economist*, vol. XI, p. 5.

TYLER, G. J. (1966) Analysis of the solution tableau of a standard linear programming problem. *Journal of Agricultural Economics*, vol. XVII, p. 2.

CHAPTER 6

A Note on the Areal Patterns in the Value of Early Potato Production in South-west Wales, 1967

J. G. TYRRELL

As the title to this paper suggests, the scope of the subject-matter is limited, both in its regional interests and in the observations it seeks to present. This is because the regional information contained within it is based on a number of questionnaires, a little over 50 per cent of which have yet to be returned at the time of writing (March, 1968). Despite the fact that no detailed analyses can be presented here, it is nevertheless possible to indicate the range of environments experienced in the early potato producing areas of Pembrokeshire in south-west Wales and to suggest the extent to which various environmental factors can be expressed in financial terms.

If a weather input is irregular in its occurrence and if an enterprise is dependent upon it to any degree, then it can be termed a "hazard" when an economic assessment of it is made. Where weather events are predictable they have no economic significance in that economic fluctuations would be independent of them. Differences in the probabilities of weather events occurring between areas of the earth's surface mean that important contrasts in the weather-enterprise chain effect are readily apparent on a macro-scale. But differences on a micro-scale (i.e. within a locality of a few square miles) are equally important, indeed more important to the decision-maker (in this case the farmer), particularly with a weather-sensitive enterprise.

McQuigg and Doll (1961) have shown that to analyse the effects of a weather variable upon production three sets of information are required: (1) knowledge of the specific weather input affecting the process; (2) the cost and weather functions associated with this weather input; and (3) the maximum and minimum quantities of that input which identify the profitable limits of an enterprise.

The first set of information for early potato production can be obtained from the numerous laboratory and field experiments and observations that have been carried out by workers concerned with the growth processes of the potato. Although it is possible to isolate the more important weather inputs in this way it is not so easy to recognize the significance of different intensities of input in terms of production variations, since the conditions of experiments are frequently so variable. Nevertheless important progress is being made in this field.

A summary of some of the main results of this work is shown diagrammatically in Fig. 23. There are three main stages in the growth of the potato in the field. Each of these is characterized by different growth processes, requiring differing environmental conditions. The pre-planting stage, concerned with storage and sprouting does not concern us here, other than to note that the greatest measure of environmental control is possible at this stage at present. The other stages are pre-emergence (between planting and emergence), a phase dominated by haulm growth, followed by tuber initiation and the dominance of tuber growth.

Between planting and emergence the most favourable conditions for early growth appear to be low temperatures, since the time to emergence tends to increase at about the rate of one day per degree Fahrenheit above 56°F (13·3°C) (Borah and Milnthorpe, 1959). It is particularly important to plant physiologically old seed for early growth (Headford, 1962; Toosey 1963). The lower limit to soil temperatures is when frost damage occurs. But the identification of the critical temperature is by no means certain. Water requirements at this stage depend largely upon the degree of plant growth, and is therefore related to temperature conditions (Letnes, 1958). But Penman (1962) has demonstrated a very close correlation between growth (measured in terms of yield) and total potential transpiration, the latter being adjusted to exclude periods when the soil moisture deficit was below an assumed critical value of 1 in. This can be taken as a guide to moisture requirements throughout the life of the plant.

During the period of dominant haulm growth, temperatures in the range of 56–68°F (13·3–20·0°C), with a high level of radiation seem to give optimum conditions for hastening tuber initiation. Although haulm

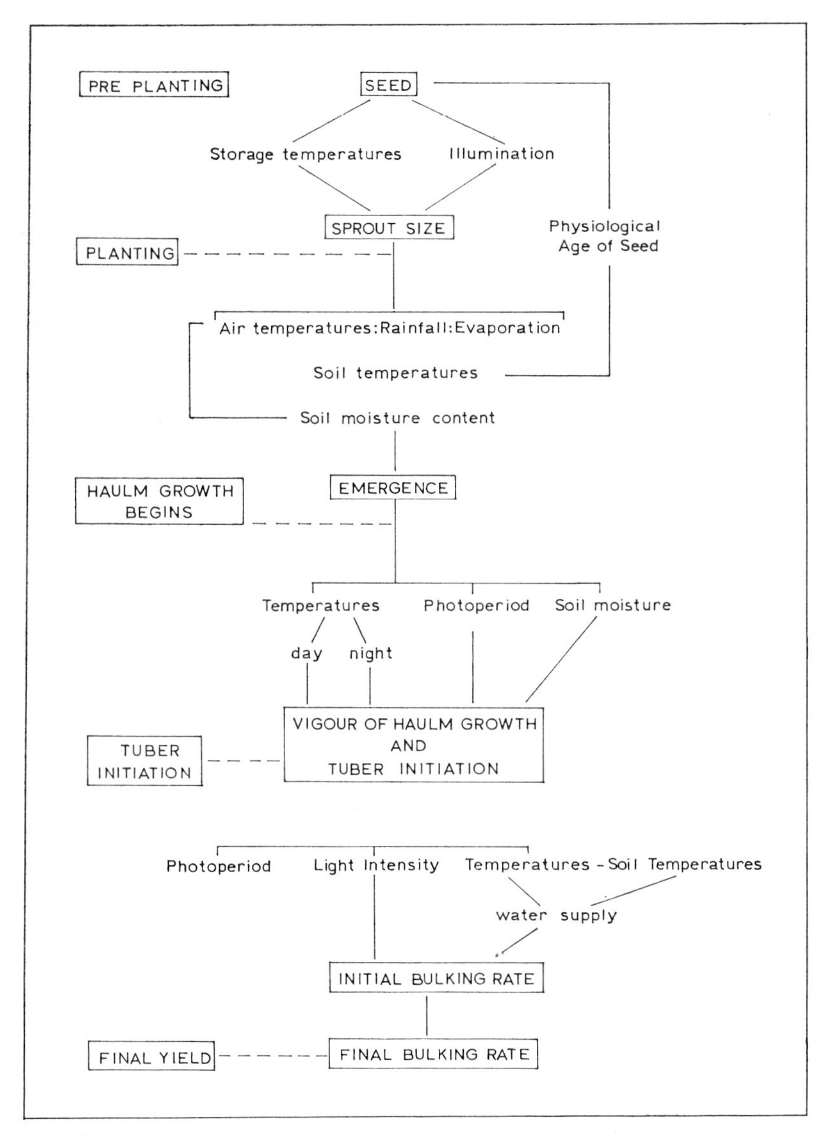

FIG. 23. Weather factors in the growth stages of the potato plant. (Adapted from McQuigg and Doll.)

growth is less vigorous than with higher temperatures, growth does not extend over such a long period, leaves are larger and auxiliary growth is less (Borah and Milnthorpe, 1959; Bodlaender 1960, 1963; Gregory, 1954).

This intermediate range of temperature is again very favourable for tuber growth. High temperatures are important since low night temperatures produce a large number of tubers, bulking relatively slowly; but high night temperatures decrease tuber yield (Bodlaender, 1960; Gregory, 1954; Went, 1957). Daytime soil temperatures of 59°–64°F (15·0–17·8°C) seem to be optimal, falling to around 50°F (10·0°C) at night (Jones *et al.*, 1922). Particularly critical during this phase is the period of 10–14 days following tuber initiation when tuber weight increases exponentially with time. A longer period follows to harvesting when the increase in tuber weight is linear with time. For a crop such as the Pembrokeshire early potato, planted in mid-February and early March, this 10–14 day period

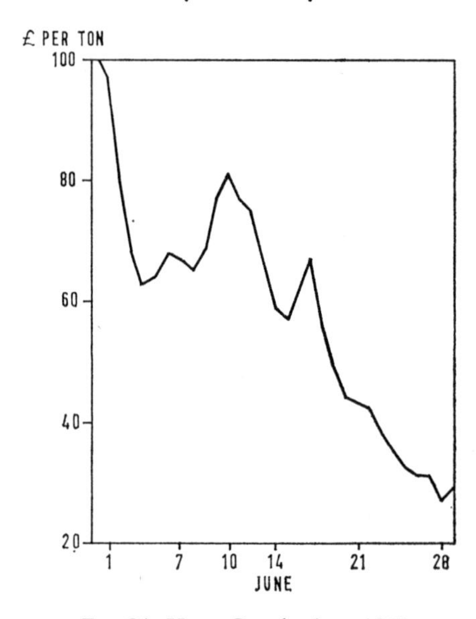

Fig. 24. *Home Guard* prices, 1967.

would occur at the end of April and the beginning of May. Freedom from weather hazards is therefore particularly vital at this time.

The second and third sets of information referred to by McQuigg and Doll (1961) can be indicated by data supplied by Pembrokeshire farmers themselves, if the depression in market price resulting from later lifting or lower yields is regarded as the economic cost of the weather input(s) responsible for the delay in marketing or growth. There is normally considered to be an inverse relationship between delay in lifting and market price. This tends to be true, but Fig. 24 shows the "double peak" in market prices. This tends to blur ecological contrasts when expressed in

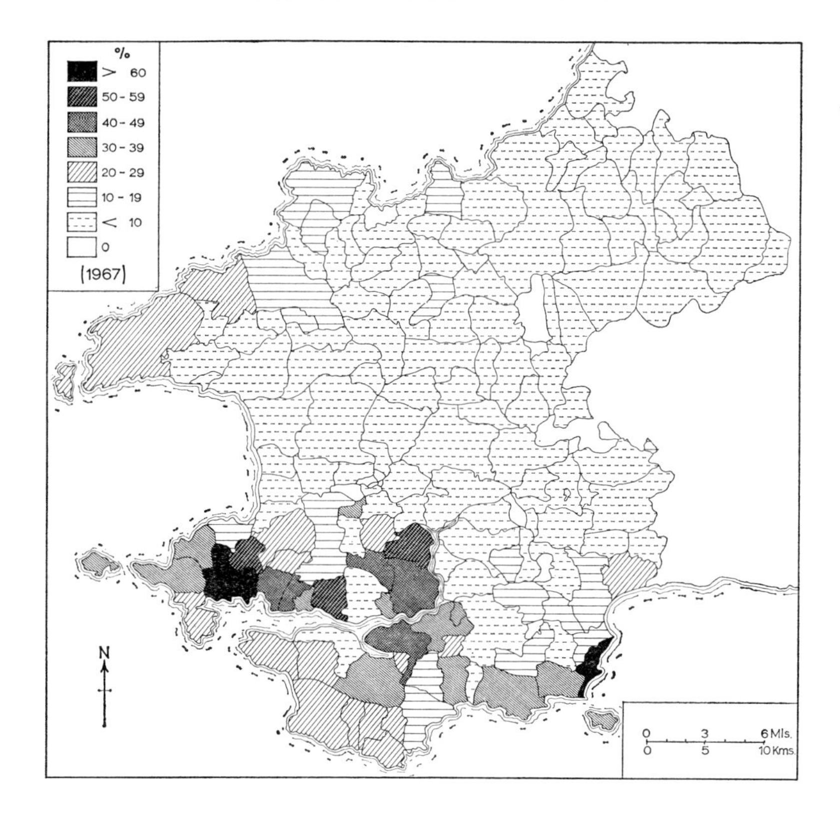

FIG. 25. West Pembrokeshire: early potato acreages by parish (percentages of total arable land), 1967. (Source of data: June 4th Returns of the M.A.F.F., 1967.) (Map research by Mrs. A. Timmis.)

economic terms. A slight delay could therefore be profitable. But the general characteristic of the price curve holds true, namely that price falls over time.

The main features of the distribution of the crop are seen in Fig. 25. Two areas stand out, the St. David's Peninsula (particularly the northern coastal zone) and the southern quarter of the county between Haverfordwest and the coast. The distribution of gross margin per acre (Fig. 26) depicts the areal pattern of economic returns.† A map of this nature is useful to indicate the range of contrasts found within the area of production and to ascertain the limits of profitable production. The same two areas stand as before. This map includes many diverse factors, some of which are

† This map is based on a survey carried out by D. Theophilus, of the N.A.A.S. in 1967.

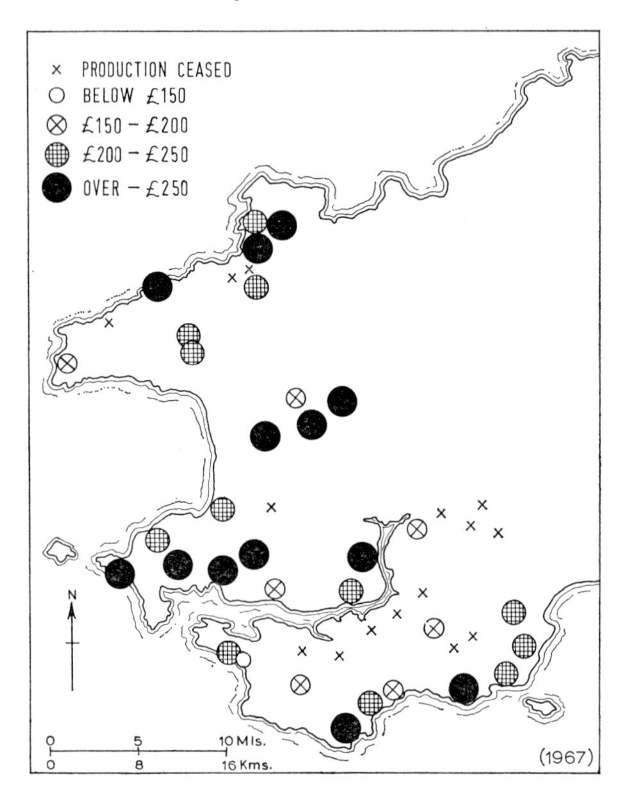

FIG. 26. West Pembrokeshire: early potato production on selected farms; gross margin per acre. (Based on a survey made by D. Theophilus of the N.A.A.S. in 1967.)

eliminated in Fig. 27 where lifting dates for one variety only (*Home Guard*) are considered, the date taken being that of the first field harvested for each farm, on the assumption that this would be the earliest field in terms of growth and yield. Most farms begin lifting at 4 tons/acre so this has been the yield taken in each case. The isopleths thus presented show a distinct gradient across southern Pembrokeshire from Broad Haven towards Tenby. A similar feature is suggested in northern Pembrokeshire between Mathry and the coastline.

So far in the discussion no environmental hazards have been suggested as explaining these distributions. For a correct appraisal of weather events contributing to crop responses, described on the maps, it would be necessary to set up agro-meteorological stations in each field for which information has been provided. Such a task is impossible. But what is not

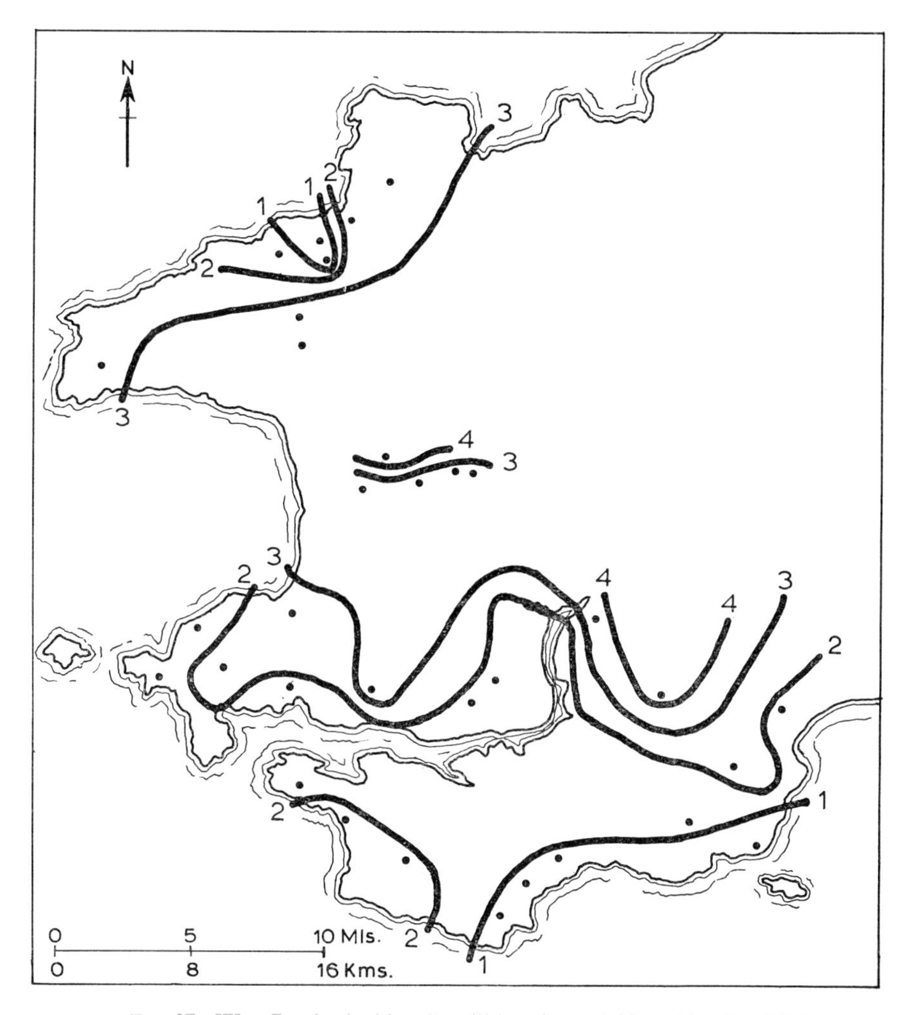

FIG. 27. West Pembrokeshire: first-lifting dates of *Home Guard* in 1967.

impossible is the use of known plant responses to give weather inputs (as described earlier) to assess agro-meteorological conditions by means of site features. That is, by knowing the relevant parameters (or weather inputs) together with distributions of physical phenomena (which by experiment and observation are known to produce different levels of weather inputs), areal variations in production can be explained (or predicted, etc.)

In Fig. 27 some of these physical phenomena stand out. The change of soil from sandy loams of the Old Red Sandstone to the heavier soils further north produces an effective delay in growth of up to 2 weeks. In financial terms this would be interpreted in different ways from year to year. But the contrast would always be there. In 1967, based on average market prices the difference would have been £69 per ton compared with £55 per ton—not very large owing to the double peak in market prices in that year (Fig. 24). Such a difference is obviously not prohibitive in production.

There is usually a deterioration of growing conditions inland from the sea. This again is mainly an effect of temperature. A coastal site may gain

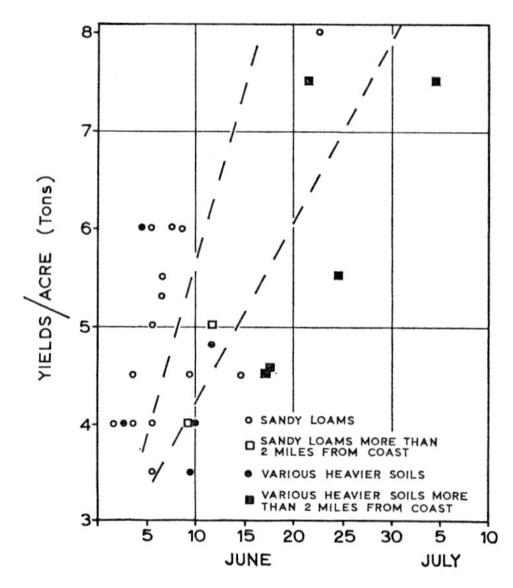

FIG. 28. West Pembrokeshire: lifting dates and yields of the earliest field per farm. (The sample of farms used is identical with the one used for Fig. 27.)

up to a week from one located 2 miles inland. This had no added advantage in 1967 between the first and second weeks of June, but between the end of the second and third weeks it made a difference of £24 per ton (e.g. near Little Haven). This effect is very marked along the north coast between St. David's and Fishguard where lifting dates differ by 4 weeks, and the average market price by £45 per acre over a very short distance.

Figure 28 incorporates lifting dates and yields for the same farms used in Fig. 27. This suggests bulking rates differ in different areas of the county, a factor highly significant in determining yields. Although factors other than soil and the distance from the coast are involved, these do stand out again, the coarser textured sandy loams and areas within 2 miles of the sea having about a week's advantage over heavier soils and inland fields. It is interesting to note again that heavier soils near the coast can be very profitable. This shows the very strong influence of the coastline upon the earliness of growth.

The frost hazard is the negative weather input usually assumed to be responsible for the restriction of the early potato to the south and north coastal margins. Figure 29 shows the increase in the number and intensity of the frosts inland. If this is true of frosts, Fig. 29 would indicate that 3 to 4 miles from the coast is as far as production can be profitably extended. Yet production in 1967, when the observations in this table were taken, was to be found as far inland as the third site in the table, indeed on the very same farm. Despite the frosts, farms located 3 and more miles inland, respectively, were both lifting in the second week of June at 4 tons per acre. The risk involved in producing weather-sensitive crops in such an area is seen, however, in a comparison between 1964 and 1965 at a farm about 4 miles north of Haverfordwest and 7 miles from the sea. In 1964, lifting between June 20th and July 3rd, a yield of $8\frac{1}{2}$ tons per acre was obtained. In 1965 in the same field, planting the same variety, 18 air frosts were recorded at the nearest meteorological station (Haverfordwest) between March and the end of May, 8 occurring during the last two months when none were recorded at the other standard meteorological sites in Pembrokeshire, nearer the coast. Lifting was as late as July 14th to 19th, and a yield of only 4 tons per acre was harvested. Radley (1963) noted in an experiment that a frost temperature as low as 17°F (−8·3°C) at the ground surface, severely damaging the foliage, resulted in a comparatively small reduction in final yield. There was a rapid recovery after relative growth rates had been reduced for approximately 2 weeks. However, as the early potato crop is not left to grow to maturity it is likely that such frosts would have had a greater effect on yields realized at lifting than otherwise.

There are a number of locational characteristics which contribute to areal contrasts in temperature. Moisture conditions have been shown to be equally relevant. Based on potential transpiration values, Fig. 30 has been constructed to show the total irrigation that would have been required to July 1st, 1967, assuming irrigation is applied at the end of a 2-week period if the soil moisture deficit was greater than 1 in. This shows the need to apply irrigation increase towards the south-west coast. The contrasts shown are basically due to rainfall differences, with rainfall increasing inland towards the Prescelly Mountains. Assuming water for

Sites	Minimum ground surface temperatures (°F) (frosts per week in brackets)				Minimum 8 in. air temperatures (°F) (frosts per week in brackets)			
	Coast	3 miles	9 miles	12 miles	Coast	3 miles	9 miles	12 miles
Date								
Mar. 4–11	35·5	33·0	32·0 (3)	30·5 (2)	36·6	36·0	34·5	31·0 (1)
12–18	36·0	32·0 (1)	28·5 (3)	26·0 (4)	36·3	33·7	31·5 (4)	N.A.
19–25	33·5	31·5 (1)	29·0 (2)	27·0 (5)	37·2	33·0	30·0 (2)	28·0 (3)
25– 1 Apr.	25·0 (2)	23·5 (5)	26·0 (6)	23·0 (7)	28·5 (2)	24·0 (4)	25·0 (4)	20·0 (7)
Apr. 2–8	30·3 (1)	28·2 (1)	30·2 (1)	31·0 (1)	32·9	29·7 (1)	33·5	32·0 (1)
9–15	33·4	29·5 (3)	30·0 (3)	31·0 (2)	36·5	30·5 (2)	32·0 (2)	28·0 (1)
16–22	32·8	31·3 (1)	31·8 (1)	28·0 (3)	36·5	31·0 (1)	31·0 (2)	27·0 (4)
23–29	38·2	33·0	31·2 (1)	31·0 (1)	39·5	32·0 (1)	33·5	30·0 (1)
30–6 May	28·0	25·7 (3)	29·0 (2)	23·0 (3)	32·6	27·0 (2)	24·5 (2)	22·0 (3)
May 7–13	39·0	38·5	38·5	35·0	41·0	34·3	33·8	36·0
14–20	32·8	35·0	34·0	31·0 (1)	36·0	35·6	34·0	31·0 (1)
21–27	41·0	41·5	42·5	41·0	43·5	41·0	42·3	41·0
28–3 June	39·0	40·0	38·5	38·0	41·6	35·5	N.A.	33·0
June 4–10	35·9	37·0	34·8	33·0	37·6	32·6	30·4	31·0 (2)
11–17	41·0	41·3	44·0	37·0	43·0	35·3	37·0	35·0
18–24	43·2	45·0	46·9	42·0	47·5	39·2	41·6	38·0
23–1 July	41·3	42·0	48·7	43·0	47·0	38·1	39·6	38·0

FIG. 29. Weekly frosts and minimum temperatures, 1967 (N.A. = Not available).

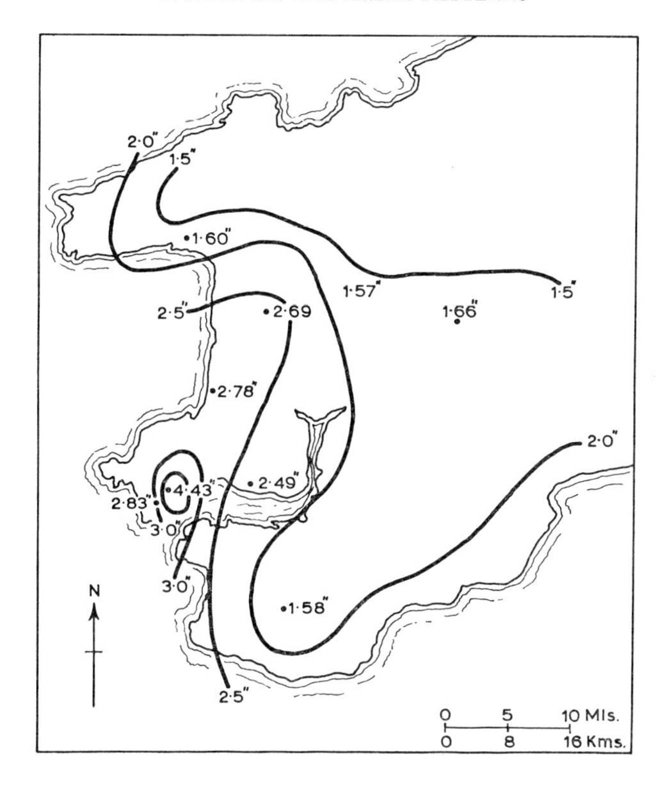

FIG. 30. West Pembrokeshire: irrigation required to July 1st, 1967.

irrigation were available from ponds, streams and wells, etc., the cost per acre would be at least £4 5*s*. for 1 in. and £5 7*s*. for 2 in., but it is probably, in fact, greater.† Without irrigation, growth in June (when these deficits developed) would have been very restricted, though many producers had lifted before irrigation became necessary.

Figure 31 introduces the problem of scale into the discussion of areal contrasts in the value of the potato crop in Pembrokeshire. Although the map of irrigation-need showed a gradual increase in the south-west, two stations in the extreme south-west afford a marked contrast, due mainly to differences in rainfall (column 3). This is surprising because of the standardization of both sites. One suspects a sheltering effect at the Dale Fort site. Nevertheless, it raises the question of local but very important variations in growth parameters, due to local site features. In this case the

† This is based on information from the Natural Resources Committee's Report on Irrigation, 1962.

	Period		Average potential transpiration	Rainfall	Deficiency	Soil moist deficit	Planned deficit	Irrigation	Sun. cor. fac.	Corrected soil deficit	Actual deficit (without irrigation)
	Apr.	1–15	0·97	1·03	—	—	1·00	Nil	—	—	—
		16–30	1·00	0·33	0·67	0·67	1·00	Nil	+0·19	0·86	0·86
Dale	May	1–15	1·40	2·04	−0·64	0·22	1·00	Nil	—	—	0·22
Farm		16–31	1·41	2·57	−1·16	—	1·00	Nil	−0·19	—	—
	June	1–15	1·75	0·01	1·74	1·74	1·00	1·74	—	—	1·74
		16–30	1·65	0·56	1·09	1·09	1·00	1·09	+0·04	0·04	2·87
	Apr.	1–15	0·97	0·44	0·53	0·53	1·00	Nil	—	—	0·53
		16–30	1·00	0·15	0·85	1·38	1·00	1·38	+0·19	0·19	1·57
	May	1–15	1·40	1·29	0·11	0·30	1·00	Nil	—	—	1·68
Dale		16–31	1·41	1·94	−0·53	—	1·00	Nil	−0·19	—	0·96
Fort	June	1–15	1·75	0·01	1·74	1·74	1·00	1·74	—	—	2·70
		16–30	1·65	0·34	1·31	1·31	1·00	1·31	+0·04	0·04	4·01

FIG. 31. Soil moisture deficits for Dale, 1967.

effect of sheltering could result in minimum irrigation costs of £7 per acre and above, to maintain a 1-in. soil moisture deficit.

Exposure and shelter affect the temperature as well as water supply. Between identical fields on south-west slopes (one is in the Marloes Peninsula on the cliff top and the other is the neighbouring parish of St. Bride's, halfway up a valley slope) a difference of 2 tons/acre was recorded. Both fields were lifted in the first week of June when £60 to £70 per ton was being realized. On the same farm at Marloes a combination of site characteristics of aspect and exposure, the former being the dominant, produced a 1½ tons/acre difference when lifted. Many other examples could be quoted. The physical damage due to wind and salt is a particular hazard associated with exposed sites along the coastline. Gale force winds are particularly frequent during the early stages of haulm development and growth.

It has been suggested, and demonstrated with specific examples, that weather inputs and their costs can be studied regionally or locally. Any distribution is a result of environmental conditions at both of these levels. Figure 32 plots lifting date and yield for five farms, each having three or more fields under the variety *Home Guard*. Yield has been plotted against lifting date for each field to give some indication of bulking rates. Each point is a product of both regional and local conditions, the latter interpreting the former. Whether the regional component is as important as Fig. 27 suggests, remains to be seen. The positions of the curves are mainly due to regional effects, but the relationship of the points in the curve to each other is due to local environmental characteristics.

Knowledge of weather inputs and their cost functions must be applied to be of value. As already suggested, this calls for considerable quantities of data concerning the distribution of the environmental parameters known to be important. Without this areal data it would be extremely difficult to explain accurately areal contrasts and trends in production. This is of great economic importance, not merely in terms of farm income, but also in appraising realistically, and developing, agricultural resources. What measurements are available from standard meteorological stations relate to special conditions designed for analysis on a national scale. The site where it is located may be far from typical of the area for which weather data is required for the purpose of, for example, planning. This is particularly true of a weather-sensitive crop such as early potatoes where slope, soil, aspect, etc., can produce considerable differences in growth potential.

A measure such as *site potential* could help to overcome the practical

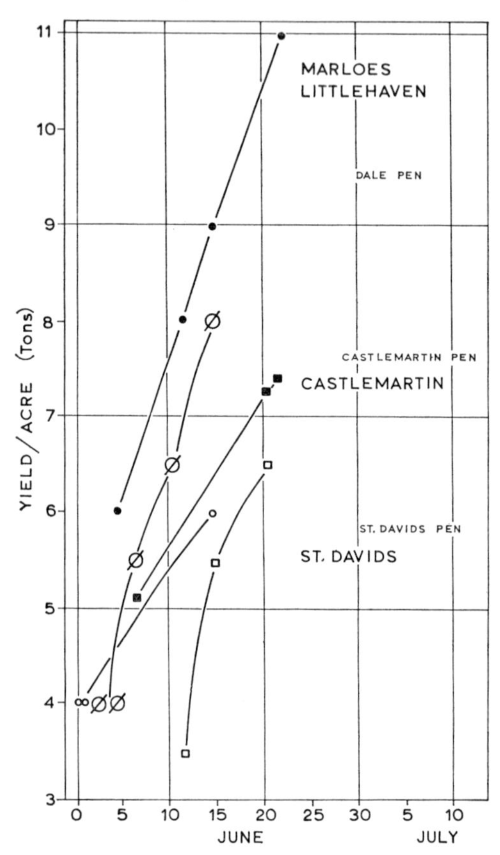

FIG. 32. Lifting dates and yields from five Pembrokeshire farms growing *Home Guard*, June–July, 1967.

problems of instrumentation and observation in measuring meteorological parameters at numerous points in a locality. The term "site" is meant to refer to all non-meteorological characteristics at a point in space, which produce differences between relatively small areas within a given weather type. Soil, slope, aspect, exposure, etc., have already been mentioned as examples. The scale on which this is applied will depend upon the enterprise being studied. A crop which is extremely weather-sensitive would need to be related to small-scale differences. The final criteria for this would be the scale relevant for producing information useful for the producers. Site potential could be measured in several ways. One suggested is as follows. Having determined the significance of a weather input, the probability of its occurrence on particular sites, compared with a standard reference site (e.g. a standard meteorological site), could be computed. If these were then classified, important indices concerning the homogeneity of an area in terms of site would be produced, with implications for particular enterprises. When various types of data and forecasts were being prepared, they would immediately have great relevance to the wide area.

A study of the agricultural significance of spatial variations in production in south-west Pembrokeshire suggests that, having acquired the three sets of information McQuigg and Doll (1961) have shown necessary for analysis of the effects of a weather variable upon production, it will not necessarily be an easy matter to apply the knowledge required. In theory, spatial considerations fall under their second and third sets of information. But since they are crucial in producing meaningful economic results and decisions, their importance cannot be over-emphasized. Indications as to the financial implications of such contrasts in site characteristics have been made, and these mean that the farmer cannot afford to be without a great deal more local information than is at present available to him.

REFERENCES

BODLAENDER, K. B. A. (1960) *Jaarb. Inst. biol. scheik. Onderz*, Wageningen, p. 69.

BODLAENDER, K. B. A. (1963) The influence of temperature, radiation and photo-period on development and yield, in Ivins, J. D. and Milnthorpe, F. L. (Editors), *The Growth of the Potato*, London, 1963.

BORAH, M. N. and MILNTHORPE, F. L. (1959) Annual Report, University of Nottingham, School of Agriculture, 1959, p. 41.

GREGORY, L. E. (1954) Some factors controlling tuber formation in the potato plant. Ph.D. thesis, University of California, Los Angeles, U.S.A.

HEADFORD, D. W. R. (1962) Sprout development and subsequent plant growth. *European Potato Journal*, vol. 5, p. 14.

JONES, L. R. *et al.* (1922) The influence of soil temperature on potato scab. *Bull. Wis. Agr. Exp. Sta.* No. 53, pp. 1–35.

LETNES, A. (1958) The effect of soil moisture on the sprouting of potatoes. *European Potato Journal*, vol. 1, p. 27.

MCQUIGG, J. D. and DOLL, J. P. (1961) *Weather Variability and Economic Analysis*, University of Missouri, Coll. Agric. Res. Bull. 771.

MINISTRY OF AGRICULTURE, FISHERIES & FOOD (1964) *The Calculation of Irrigation Need.* Technical Bulletin No. 4.

NATURAL RESOURCES (TECHNICAL) COMMITTEE (1962) *Irrigation in Great Britain.* H.M.S.O.

PENMAN, H. L. (1962) Woburn irrigation: (1) purpose, design and weather; (2) results for grass; (3) results for rotation crops. *J. Agric. Sc.*, vol. 58, pp. 343, 349 and 365.

RADLEY, R. W. (1963) The effect of season on the growth of the potato, pp. 211–19 in Ivins, J. D. and Milnthorpe, F. L. (Editors), *The Growth of the Potato*, London, 1963.

TOOSEY, R. D. (1963) The influence of sprout development on subsequent growth and yield, in Ivins, J. D. and Milnthorpe, F. L. (Editors), *The Growth of the Potato*, London, 1963.

WENT, F. W. (1957) *The Experimental Control of Plant Growth*, Chronica Botanica Co., Waltham, Mass., U.S.A.

CHAPTER 7

Weather and Risk in Forestry

P. A. WARDLE

THE January gale in Scotland, better known for the damage it did to Glasgow and neighbouring towns, also blew down in west central Scotland some 30 million cubic feet of timber—something like the volume normally cut in 18 months in the whole of Scotland. In 1967, more than 1000 million cubic feet were blown in Europe, amounting to 50 per cent of the normal cut in such countries as West Germany. Fortunately, fire losses recorded in Britain are on nothing like the same scale. In other climates and conditions forest fires reach spectacular dimensions: a bushfire in Tasmania in 1967 burned over 1000 square miles and killed 53 people, and the Bel Aire fire near Los Angeles, which destroyed 480 houses and did damage valued at 30 million dollars, is a fairly recent example in the long history of disaster in the United States.

Both windblow and forest fires are phenomena related to the weather and both are important in the economic management of forests. This paper will concentrate on the implications of hazards for the management of commercial forests. A number of features that both wind-throw and fire have in common will first be indicated; then a major difference between the two will be described; then a discussion will follow on the major features of the forestry production process in relation to these two hazards. In conclusion suggestions will be made as to how weather prediction can help management in the two cases.

THE OCCURRENCE OF WIND-THROW AND FIRE LOSSES IN FORESTRY COMMISSION FORESTS

In Figs. 33 and 34a the annual losses from wind and fire are shown. The series of statistics on wind-throw is unfortunately short, since wind-throw has appeared as a significant loss in Forestry Commission areas only in recent years. Records of fire loss are available from 1929 and earlier series are reported by Charters (1961). One must be most cautious in the interpretation of such time series, since as Davis (1965) points out, the observed results are tied up with changing levels of inputs and changes in the scope of the problem. This is particularly true in the case of these British series since the extent and composition of the forests is changing from year to year. This affects wind-throw: with the plantations being predominantly young, the risk of damage from wind increases from year to year as trees

Year ending March 31st	Acres
1962	1250
1963	350
1964	180
1965	220
1966	90
1967	340
1968	10,000

FIG. 33. Wind-throw damage to Forestry Commission plantations, 1962–8.

Year ending September 30th	Number of fires	Area burned acres
1954	1340	390
1955	2835	275
1956	2045	4080
1957	925	120
1958	1030	400
1959	5600	395
1960	4090	1595
1961	2235	305
1962	1820	990
1963	920	940
1964	1155	465
1965	865	785
1966	255	400
1967	310	300

FIG. 34(a) Fire losses in Forestry Commission plantations, 1954–67.

Year	Total area burnt acres	Number of fires over 20 acres	Approximate percentage of total area in fires over 20 acres
1962	990	9	60
1963	940	13	70
1964	465	2	50
1965	785	9	40
1966	400	5	40
1967	300	3	30

FIG. 34(b). Fire losses in Forestry Commission plantations, 1962–7.

grow older, taller and thus more susceptible to damage. The further study of these series relating losses to causes for purposes of prediction would require sophisticated analysis. Note the variation of losses in different years and the apparent skewness of the distribution. The proportion of losses by extent in each year is graphed in Figs. 35 and 36.

It will be seen from Fig. 34a that the number of fires recorded is large in most years but the average size of fire is small—only in a few years has the average size exceeded 1 acre. On the other hand a large proportion of the loss is caused by the comparatively few larger fires. In Fig. 34b losses from fires greater than 20 acres in extent in recent years (less than 1 per cent of the total number of fires) are shown to account for from 30–70 per cent of the total area damaged. Davis (1965) remarks that in California fires involving over 300 acres account for only 2 per cent of the number of fires but for 85 per cent of the area burnt. In comparison, Figs. 34a and 34b suggest the main variation in losses may be attributable in the British case to the incidence of large fires. The fact that in 1956, the year with the largest total loss in Fig. 34a, one fire burnt 1000 acres tends to support this view.

It is not possible to report wind-throw in quite the same way as fire, as its occurrence on any occasion tends to be distributed over a forest or group of forests. What one can say is that damage in any year tends to include a large proportion of the total occurring on a few dates, caused by a few severe storms, and the occurrence is concentrated in limited regions affected by those storms. An equivalent characteristic is found in the case of fire losses in California. Davis (1965) shows for 4 years reported that in 3, 60–90 per cent of the year's losses occurred on 6 or fewer days.

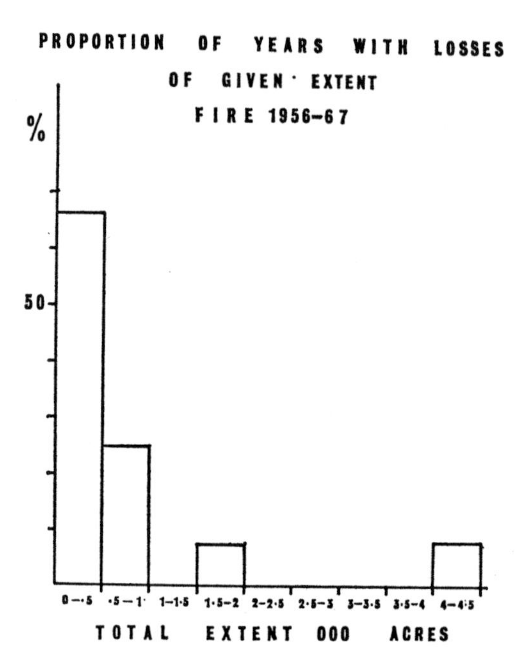

FIG. 35. Proportion of years with timber losses of given extent due to fire (1956–67).

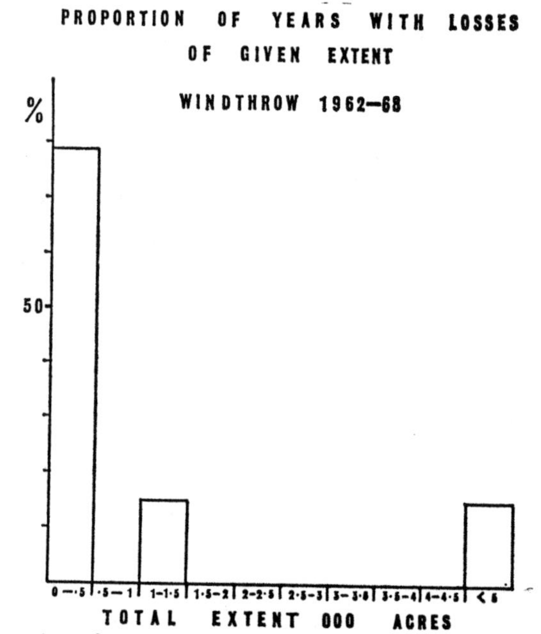

FIG. 36. Proportion of years with timber losses of given extent due to wind-throw (1962–8).

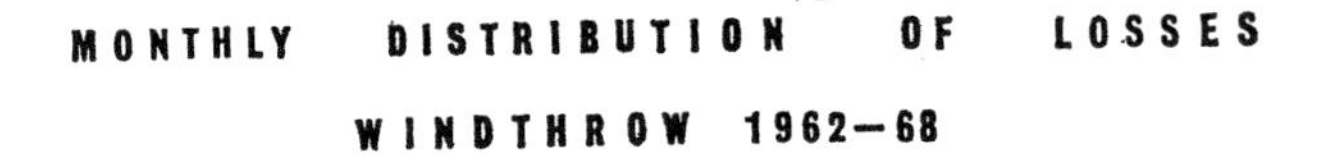

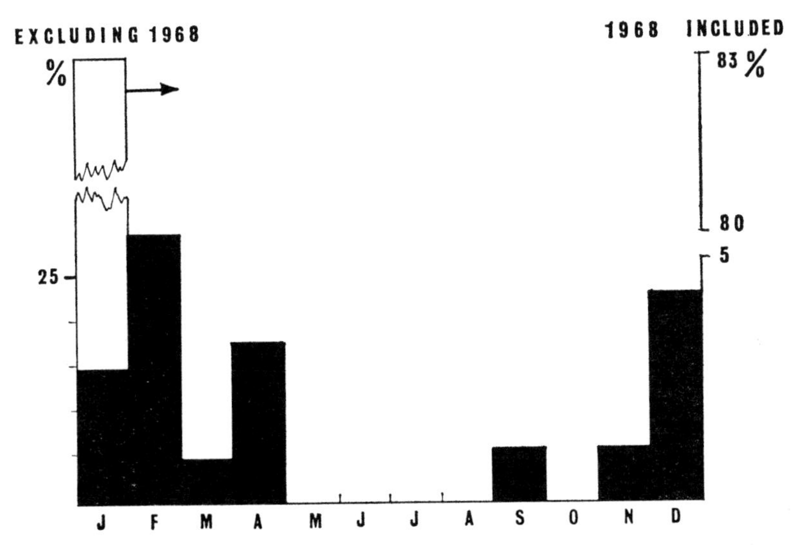

FIG. 37. Monthly distribution of timber losses due to wind-throw, (1962–8).

The seasonal distribution of incidence, both of fire and wind-throw, is worth noting. From the few years' data available on wind-throw the picture in Fig. 37 shows the seasonal occurrence. It is notable that the two major losses in recent years, 1953 and 1968, were both in January gales. Figure 34b shows the seasonal occurrence of fires up to 1956 after Charters (1961). By comparing two periods he shows that the seasonal distribution is fairly constant. Davis (1965) for California shows a similarly stable distribution of the seasonal incidence of fires. By reference to large fires only, Davis's distribution probably gives a good indication of the incidence of loss. Reference to the seasonal distribution of large fires over recent years suggests that the distribution of losses is more concentrated than the distribution of numbers of fires (Fig. 38).

In the more extreme conditions of California the variability from day to day in the incidence of fires, which is closely correlated with weather conditions, is of great significance. In this country (Britain) there is also important daily variation in the risk within seasons. Charters (1961) also shows the variation in risk during the day (see Fig. 39). One has to be

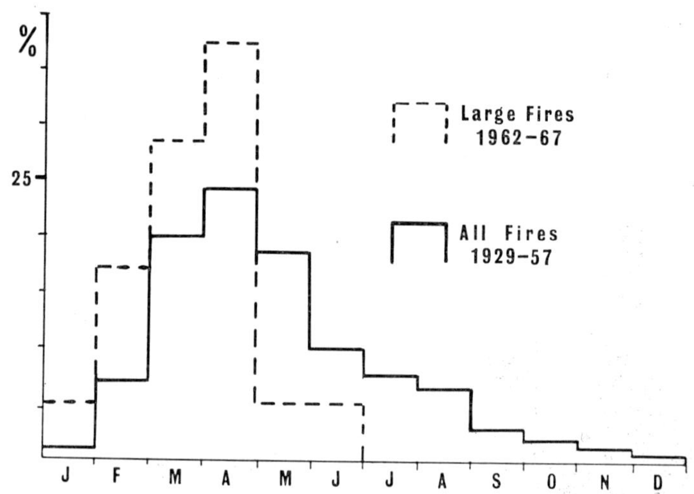

FIG. 38. Monthly distribution of timber losses due to fire: large fires, 1962–7; all fires, 1929–57.

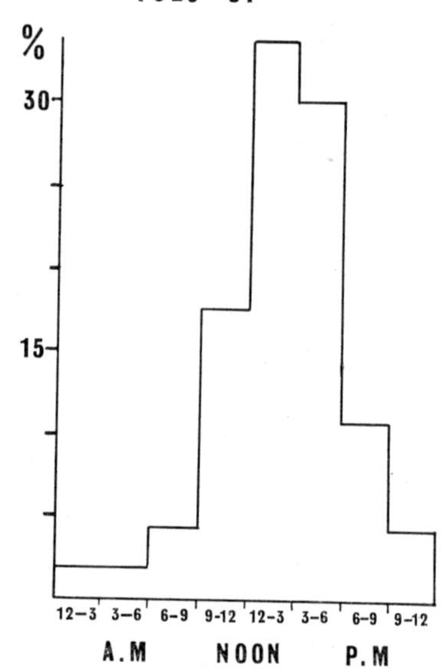

FIG. 39. Distribution of fire starts by time of day, 1929–57 (after Charters).

careful, of course, in ascribing a temporal variation exclusively to weather, since variation in human behaviour (humans in our conditions are the principal cause) may be confounded with season, weather and time of day.

REGIONAL VARIATION IN THE PATTERN

It is certainly possible to find weather-related differences in the incidence of both wind-throw and fire loss between different regions in Great Britain. The general distribution of high and low risk areas from the point of view of wind-throw may be located by inspection of isovent maps, it being perhaps reasonable in general to suppose the incidence of very high winds to be related to the average wind speed (Golding and Stodhart, 1952) (see Fig. 40). Other things being equal, the likelihood of wind-throw is greatest in the most windy places. Similarly, fire risk is related to the monthly distribution of rainfall and humidity, other things being equal, the risk being greatest in periods of low rainfall and low humidity and in those places where these conditions recur over the longest periods.

The object in this introductory discussion of the incidence of wind-throw and fire losses has been to indicate the variation from time to time

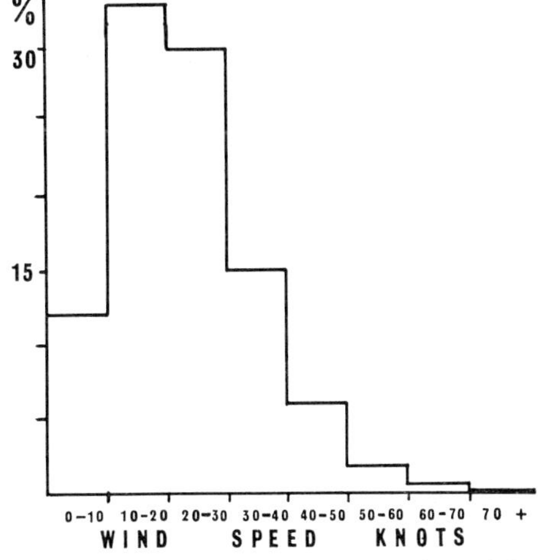

FIG. 40. Indicative distribution of wind by hourly wind speed for the year for a very windy site (after Golding and Stodhart).

in the extent of losses and to indicate the random distribution both in time and location of actual losses over the area at risk, other things being equal. It has also been indicated that the incidence of loss can generally be related to season and, more particularly, to weather within seasons and to variations in the seasonal patterns between regions.

Both wind-throw and fire loss are components of the function of forestry production and their incidence can be affected by the treatments adopted. The commercial objective of the forest manager is to carry out that combination of treatments which will make the sum of opportunity cost of treatment and loss, a minimum. In the case of fire, the opportunity cost of treatment is mainly the cost of resources used on advanced protection measures for standby and fire-fighting, and the loss is a compound of the current value of the timber destroyed, the value of the lost potential and the opportunity cost of resources used on replanting the crop now, rather than at the end of the normal productive cycle. In the case of wind-throw, the opportunity costs of treatment are the cost of resources used in draining and cultivation designed to improve stability, the opportunity costs of adopting treatments which in the absence of wind risk would not be optimum, and the loss through wind-throw includes the increased cost of handling wind-blown material compared with normally felled material, the loss in value of material due to damage when blown, the value of lost potential and the cost associated with replacing now rather than at the optimum age. These are necessarily assessed in terms of present values.

The Forest Manager, in making his decision about treatments, is faced with a system in which the actual outcome is uncertain. The question of the best course to adopt in such circumstances has been widely discussed, but it is apparent that besides searching for a course which will reduce the total cost there will also be a concern to reduce variability. Davis (1965) argues that reductions in the variance of wildfire activity would have value to society over and above a reduction in the mean. This has been shown to depend on the marginal utility of losses being other than constant (e.g. Dorfman, Samuelson and Solow, 1958). It is apparent that the disruption caused by exceptionally large fires and extremely extensive wind-throw is such as to introduce costs that would not arise in the case of more usual occurrence. In general, an important function of management is, through the collection of appropriate information, to obtain a reduction of uncertainty allowing improved deployment of resources.

It is possible to take advance action to reduce the risk of damage both from wind and from fire hazards. Special precautions can be taken on the particular days when conditions make fire risk exceptionally high; and fire-fighting affects the extent of damage from fires once started. There is no measure equivalent to fire-fighting in the case of wind-throw, the trees in the path of a particular storm will be blown if the forces are sufficient and there are no means yet known of altering particular winds.

The nature of the wind and fire risk in the forestry production cycle and the possibility of modifying these risks through treatment will now be briefly described.

WIND-THROW

The risk of a tree being thrown by a wind of given force depends on the tree size and the tenacity of its root system, the latter being closely related to the soil conditions. Fraser (1965) indicates the general nature of this relationship for four of the principal soil types occurring in the British forests (Fig. 41). The incidence of winds of critical speed is related to the location and detailed topography. The exposure of a particular tree to critical force is affected by structure of the forest canopy. The minimum

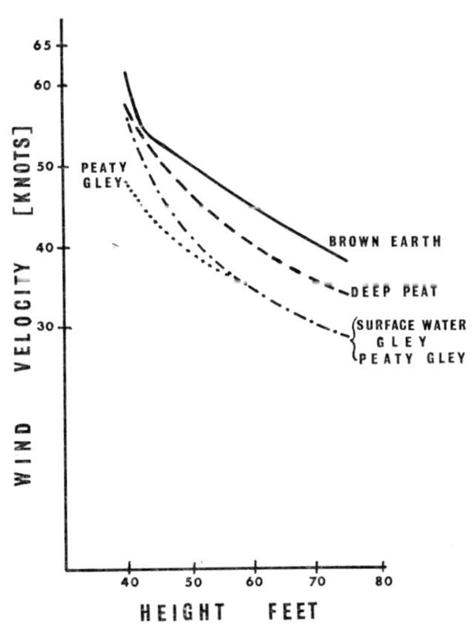

Fig. 41. The relationship between critical wind speed and tree height for four principal soil types (after Fraser).

force will be obtained with a smooth canopy; irregularities such as those caused by gaps formed in felling crops or by roads, or by thinning crops, increase turbulence and the forces to which the remaining trees are exposed.

Three main types of treatment are available which affect the incidence of wind-throw. The first of these is treatment at the time of establishing the crop; drainage and cultivation are expected to improve the soil condition from the point of view of root development and this increases the stability of the crop. Fraser's graph (Fig. 41) suggests the maximum improvement in stability obtainable by this means is with 40-knot winds, an increase from 50 ft to 70 ft in the case of gleys or peaty gleys. It is possible that so complete a transformation cannot be expected. Because, however, heights of 45–55 ft are associated with high increment of volume and value, even quite small increases in height attained are well worth achieving (Figs. 42 and 43).

Recent surveys by Pyatt (1967) have classified the extent of sites where drainage is important to stability. On certain of these young forests no drainage exists and questions arise as to whether the crops should be drained, and also whether available equipment should be concentrated in planted areas where drainage, if postponed, will either be less effective or become prohibitively expensive. Figure 43 shows that if the decline in response is in the ratio indicated in the height column, then drainage operations up to about 10 years after initial establishment are equally worthwhile. Beyond this age the combination of increasing cost and reduced response make it less worthwhile. The use of effective ploughing for draining may be expected to reduce the cost, making the smaller responses indicated in the final columns also worth achieving (see also Fig. 44).

Reference has already been made to the increase in roughness resulting from thinning: this may increase the risk of wind-throw. By producing earlier yields and increasing the value of the retained crop, thinning increases the total value of returns. Figure 45 shows that this increase in value is at a maximum when the crop is grown to complete rotation age of about 60 years in the example. If the rotation is terminated earlier the addition to value resulting from thinning is reduced. If the rotation is terminated before the crop has reached 60 ft, then this addition is quite small. If the crop is expected to be blown at 60 ft or less, the additional expenditure, namely, on early investment in roads (and possible road networks denser than the optimum for clear felling) is not likely to be justified by the additional value obtained from thinning.

Supposing the road network formed for the extraction of thinning

Sitka spruce: Yield Class 140, *discount rate* 5%

Top height, ft	40	50	60	70	80
Approximate age, years	30	35	40	50	60
Loss in D.R.[1] per acre	50	30	15	5	0

[1] D.R. values at start of rotation.

FIG. 42. Loss in present value (D.R.) when rotation is terminated at the indicated top height compared with optimum rotation.

Sitka spruce: Yield Class 140, *discount rate* 5%

Age of crop to be drained	Top height[1] to which drained crop grows (ft)	Expend-[2] iture on drainage	Gain in[3] D.R. through drainage	Net gain[4]	Approximate gain[5] in present value from increase in height of	
					5 ft	10 ft
		£ per acre				
0	70	10	39	29	8	18
5	67	14	43	29	9	20
10	62	16	44	28	12	25
15	55	22	34	12	15	30
20	47	20	13	−7	20	35

[1] The expected height to which undrained crops are expected to grow is 45 ft.

[2] In some circumstances adequate drains can be cut with ploughs at appreciably lower cost.

[3] Increase in present value compared with no draining valued in year drainage is carried out. It is assumed that the benefit from drains will last to the subsequent rotation.

[4] To year in which draining is carried out. Discounting to a common year (say, 0) shows that the maximum gain is achieved by draining in year 0.

[5] To year in which drainage is carried out. Height increase from 45 ft.

FIG. 43. Maximum value of investment in drainage, and expenditure on drainage, from 5 ft and 10 ft increases in height, in crops of different ages.

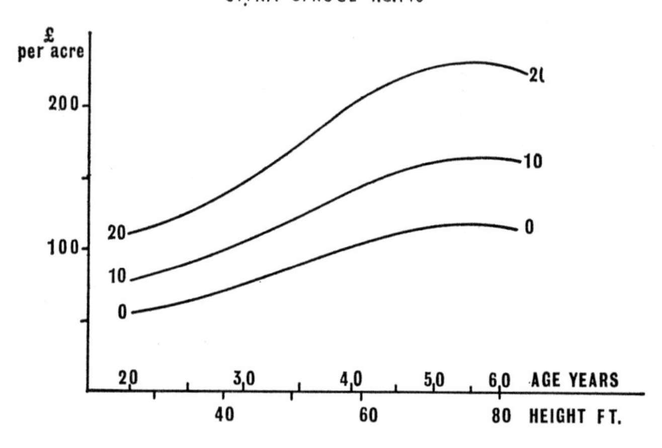

FIG. 44. Present value of returns when the forest crop is grown to various ages, viz. 0, 10 and 20 years (Sitka spruce).

Sitka spruce: Yield Class 140, *discount rate* 5%

Age, years	Top height, feet	Present value of returns £ per acre		
		Thinned	Unthinned	Difference
30	43	51	55	—4
35	51	75	72	3
40	58	93	85	8
45	65	106	92	14
50	71	114	93	21
55	76	119	91	28
60	80	119	86	33

FIG. 45. Returns from thinned and unthinned crops. (Returns discounted at the beginning of the rotation.)

Note: In the tables showing present value of returns (D.R.), uniform assumptions about prices by size of tree and price through time have been made. It would be appropriate to examine the sensitivity to variation in prices as well as to variation in physical outcome but this is not attempted here.

when the crop is 25 years of age has a discounted cost of £10 per acre, the same network formed at 35 years costs £6 per acre or at 40 years £5 per acre. It will be seen from Fig. 45 that the benefit from thinnings up to the age of 35 in this example amounts to less than the saving if roading is delayed to that age, while thinnings up to 40 years may just pay for the extra investment. If it is expected, in the absence of thinning, that the crop will stand longer it can be seen that prolonging the rotation by 5 years up to 45 years of age is sufficient to justify no thinning by itself. The additional merit of postponing roading investment would make this the most attractive course, where wind-throw is expected before 45 years of age in this example. It has to be remembered, however, that roads will be required to remove wind-blown material and postponing road construction increases the risk of wind-throw occurring before the necessary road network for harvesting operations has been constructed.

So far the discussion has referred to treatments which change the expected profitability in the face of wind-throw, either delaying its occurrence or improving the net return given an expected age at which wind-throw will occur. Costs are associated with wind-throw beyond simply the loss in potential. Damage is done to the timber reducing the value for sale, the cost of handling wind-thrown material may be greater than that in controlled harvesting; upturned roots may make preparation for reafforestation more expensive. The disorganization of orderly marketing caused by the sudden increase in supply is also likely to have its costs, requiring short-term distortion in the deployment of harvesting resources. It may also lead to raw material being utilized in lower value manufacture than is usual because the supply is out of balance with particular industrial capacities. Such costs may be avoided by reducing the incidence of wind-throw. This can be done only by harvesting the trees before they are blown. Where there is considered to be a risk of very substantial wind-throw resulting from one storm, as has been thought to be possible in the concentration of Forestry Commission plantations on the English/Scottish borders, the effect of various policies of advance felling has been examined through simulation. In this case the possible loss in potential through premature felling has to be set against losses in value through wind-throw and losses through wind-throw of varying extent. This problem also confronts the local manager who has to set the scale economies of reafforesting complete blocks against the possible loss in potential through felling prematurely the standing remains of a partially blown crop.

FIRE

The risk of fire damage is related to the occurrence of fire starts, the existence of fuel and the condition of the fuel. Fire starts are in the one case of lightning actually caused by weather, but lightning is a rare cause of fires in this country. The important causes of fire in Britain in recent years have been railways, fires spreading from neighbouring land, and persons—the most important cause, in fact, being people either starting fires intentionally, which get out of control, or inadvertently doing so. To a large extent their actions are related to the weather; "controlled" burning is carried out in conditions which allow burning, and there are more people about to start fires in fine weather.

Protection measures directed to reducing losses from fire may take several main forms. For these the following terms will be adopted: *prevention, suppression, standby* and *fire-fighting*. By *prevention* is meant such activities as education and publicity intended to persuade the public not to start fires. Among pre-*suppression* methods is included (a) the formation of non-inflammable barriers on the boundaries of, and within, areas at risk to prevent spread, (b) the provision of access routes and of installations, such as reservoirs, fire stations and fire-fighting and communications equipment. Staff training would be included here. By *standby* is meant the employment of fire-fighting crews and lookouts at times of high risk. Finally, *fire fighting* means the employment of men and equipment in fighting fires once started. Though these divisions are somewhat arbitrary they have the important distinction that the prevention and pre-suppression measures have to be carried out well in advance on the basis of the expected development of conditions, while standby effort can be deployed in accordance with the conditions that turn up, and fire-fighting is related entirely to the starts which occur. Clearly, fire-fighting will be performed with a degree of efficiency related to the effort on pre-suppression and standby. Where the incidence of fire starts is small, one would hope to increase the proportion of effort on standby and fire-fighting compared with pre-suppression. This will be achieved if the conditions (when fire starts must be expected to develop into serious fires) can be predicted accurately.

The Forestry Commission, following other fire protection organizations, has developed a method of estimating the fire danger in particular locations from day to day (Rouse, 1959 and 1961).

For given fuel types the danger rating is assessed on the basis of observations of rainfall, atmospheric pressure, humidity, temperature and wind, together with forecasts of these quantities. The manning of the lookout and standby organization for the forest is decided on the basis of these observations, seasoned with local knowledge of conditions and the behaviour of the population.

Fires are reported in detail in the Forestry Commission reports and it is hoped by analysis of the records that it may eventually be possible to obtain a more precise indication of the fire risk in relation to weather conditions and to revise guide lines on the most efficient deployment of resources for fire protection from region to region.

CONCLUSION

The main difference between the hazard of wind-throw and that of fire lies in the fact that for the first the weather is the direct causal agent, whereas for the second the weather is merely an environmental factor. If a wind producing a certain critical force impinges on a tree growing in certain critical conditions, the tree will be thrown. The conditions may be changed; the incidence of the wind, however, cannot be affected. The important information to the manager in this case is the probable incidence of winds of given speed in a particular location, so that he can decide long in advance what treatments to adopt. Both the occurrence of forest fires and their spread can be controlled; the question is what is the most efficient form and degree of control? Knowledge of weather patterns will indicate the seriousness of the fire hazard, given the existence of fire starts, in a given region, and indicate the importance that should be attached to fire protection and the likely value of pre-suppression measures. Knowledge of the day-to-day development of weather in a region allows adjustment of the lookout and standby organization from day to day.

In both cases the manager, in making decisions, will be faced with the probability that whatever choice he makes will not be optimum in the circumstances. His precautions will have been either excessive or inadequate. A clear picture of the distribution of weather hazards and their relationship to his production process will increase the likelihood that he will select appropriate long-term measures. Good forecasting will allow improved deployment of resources in the case of short-term measures.

REFERENCES

CHARTERS, H. (1961) Fires in State Forests in the years 1929–56. *Forestry Commission Forest Record*, No. 45.

DAVIS, L. S. (1965) The economics of wildfire protection with emphasis on fuel break systems. State of California Resources Agency, Division of Forestry.

DORFMAN, R., SAMUELSON, P. and SOLOW, R. M. (1958) *Linear Programming and Economic Analysis*. McGraw Hill. (See particularly Appendix A.)

FRASER, A. I. (1965) The uncertainties of wind damage in forest management. *Irish Forestry*, vol. 22, No. 1, pp. 23–30, and also see the Appendix of Fraser and Gardiner (1967) cited below.

FRASER, A. I. and GARDINER, J. B. H. (1967) Rooting and stability in Sitka spruce. *Forestry Commission Bulletin* No. 40.

FRASER, A. I. and PYATT, D. G. (1966) Crop stability measurements in man-made forests. *6th World Forestry Congress*, Madrid.

GOLDING, E. W. and STODHART, A. H. (1952) The selection and characteristics of wind power sites. British Electrical Research Association Technical Report C/T 108.

PYATT, D. G. (1967) The soil and windthrow surveys of Newcastleton Forest, Roxburghshire. *Journal of Royal Scottish Forestry Society*, vol. 20, No. 3.

ROUSE, G. D. (1959) Forest fire danger table for Southern England. *Forestry*, vol. 32 (2).

ROUSE, G. D. (1961) Some effects of rainfall on tree growth and forest fires. *Weather*, vol. 16 (9), pp. 304–11.

CHAPTER 8

The Use of Cost/benefit Studies in the Interpretation of Probability Forecasts for Agriculture and Industry: an Operational Example

E. T. Stringer

For the past 25 years, Edgbaston Observatory has been responsible for advising local authorities, industrial and commercial undertakings and, to a lesser extent, farmers in the Midlands concerning weather-dependent decisions in their spheres of operation. Since the Observatory was taken over by the University of Birmingham, the aim has been to increase the objectivity of its advisory services and, as befits a geographically-oriented establishment, to concentrate upon the solution of real weather problems in actual situations. Discussions with clients over many years have made clear the fact that every problem the Observatory has to deal with is a problem in economics and geography as much as in meteorology, and psychological factors are also important. We therefore believe that our experience is of interest in current discussions concerning the economics of weather, in that for a long time we have, in effect, been making decisions for organizations and individuals who have a weather problem. We have not, therefore, unlike many meteorologists and economists, been acting merely as passive advisers.

Our experience has impressed upon us the importance of many interesting problems which rarely appear in the literature, but only three of these will be discussed here. These are:

(a) problems involved in the practical implementation of a logical framework for decision-making;
(b) practical problems involved in the presentation of probability forecasts which are designed on the basis of cost-benefit studies;
(c) the overwhelming importance, on occasions, of "geographical accidents" in the generation of excessive weather damage to local economies in what otherwise are meteorologically unexceptional conditions.

Some years ago we decided that the basis of our future services was to be provided by the probability forecast, prepared on an objective basis. Much of the early work of our forecasters was essentially subjective, based on experience and a local feeling for the weather. Our intention has been to rationalize these hunches, by means of objective prediction techniques (Gringorten, 1959), and to discard those for which no rational basis exists. Experience has shown that hunches still have their value, and the process of producing an objective forecast, involving a study of past cases, is an excellent developer for rational intuition. In essence, therefore, our forecasts are climatological inferences, based on past records and to a certain extent on theoretical meteorology, which are sharpened when a forecast is actually made by reference to the current synoptic situation.

After the introduction of our probability forecasts, two problems arose. Firstly, there was disbelief in the probability forecast, by a wide variety of clients, in particular those without acquaintance of operational research, organization and methods, or systems analysis techniques. This attitude largely stems from the fact that the probability forecast has an impersonal, "machine-made" appearance; the attitude will be with us for many years to come. Secondly, there was the need for a logical way of making a decision on the basis of the probability forecast. The important point is that the probability of a weather event, however important this event, will mean different things to the decision-maker at different stages in his operation, and the same climatological or forecast probability has a different significance for different operations proceeding simultaneously. We decided to adopt the cost/loss ratio (Thompson and Brier, 1955) as the most suitable way out of this difficulty.

As originally developed by Thompson and Brier, the cost/loss ratio applies to situations in which (a) the effect of adverse weather on the operation, and the cost of taking action to avoid weather damage, is known in precise terms such as monetary values or man-hours; (b) the decision-maker's dissatisfaction with a loss is a linear function of the monetary value of the loss; and (c) the probability of occurrence of adverse weather is known precisely. In such a situation, if C denotes the cost of taking action to avoid weather damage, L is the cost of the damage incurred if no action is taken and the adverse weather occurs, and p is the probability of the adverse weather, then the decision rule is simple:

If $p > C/L$, it will pay to take action.
If $p < C/L$, it will pay not to take action.
If $p = C/L$, it is immaterial whether or not action is taken.

It should be noted that this rule applies only if $0 < C/L < 1$, for if $C/L \geqslant 1$, no decision situation exists and it will pay not to take protective action against adverse weather, no matter what the value of p.

The original concept of the C/L ratio has been refined to some extent by Murphy (1966), who introduced the concept of "kernel expense" to the problem. However, this concept does not change the decision rule, because if $KE(P)$ denotes the kernel cost of protecting and $KE(DNP)$ the kernel cost of not taking action and then suffering the effects of adverse weather:

$$KE(P) = pC + (1 - p)C + C$$
$$KE(DNP) = pL$$

The decision rule is expressed thus:

Take action if $KE(P) < KB(DNP)$, i.e. if $p > C/L$.
Do not take action if $KE(P) > KE(DNP)$, i.e. if $p < C/L$.

An obvious improvement to the original concept is to make use of "utilities" instead of monetary figures, since the latter are often difficult to obtain. In general, it appears reasonable to assume that the curve of monetary values against utility is hyperbolic: i.e. increasing monetary value associated with a series of decisions will always be associated with increasing utility (or with increasing "satisfaction") by the decision-maker, but the rate of increase will decrease with higher utility values. As demonstrated by Shorr (1966), if complete satisfaction is reached at a utility value S, and complete dissatisfaction (e.g. financial ruin) at a monetary value R, then:

$$\frac{u(C)}{u(L)} = \left(\frac{C}{L}\right)\left(\frac{1 - L/R}{1 - C/R}\right)$$

where $u(C)$ and $u(L)$ are respectively the values of C and L as expressed in terms of utility instead of in monetary terms. An objective decision in this case is RQ which represents the value of L corresponding to financial disaster to the operator concerned. Thus the introduction of the utility concept into the problems provides a correction factor which must be applied to the C/L ratio. This correction factor always brings about a reduction in the value of the critical probability pc ($pc = C/L$): this reduction is small if R is many times greater than L, but it may be large if R is not much greater than L.

A monetary expense scale is a positive sçale, and by equating this with a negative utility scale, it is possible (Murphy, 1966) to introduce a decision rule thus: select the action which minimizes the kernel expense, i.e. select the action that maximizes the kernel utility KU by:

$$KU = (1 - C/L)d\,(p, C/L) + \delta d(C/L, p)$$

where $d(p, C/L)$ has the value 1 if $p \geqslant C/L$, and the value 0 if $p < C/L$; and δ equals 1 if no adverse weather occurs, and 0 if adverse weather obtains. The kernel utility on "protect" occasions, $KU(P)$, thus reduces to $1 - C/L$; and the kernel utility on "do not protect" occasions, $KU\,(DNP)$, to $1 - p$. The statistical expectation of the average value of KU for a number of predictions may also be determined (Murphy, 1966), and the relative usefulness of probability predictions as against climatological (i.e. long-period) averages, as a basis for decision-making, may be determined by means of the probability score (Brier, 1950; Epstein and Murphy, 1965).

In cases for which monetary or other figures exist, the C/L ratio and its refinements provide an excellent tool, not only for the interpretation of day-to-day weather forecasts in a manner which is meaningful to the operator, but also for historical studies designed to reveal the exact role of weather as an economic hazard in the past. A classic case is the raisin industry of the San Joaquin Valley of Central California. An application of the C/L ratio to historical data for this industry demonstrated (Kolb and Rapp, 1962) its climatic basis, and provided realistic estimates of the financial cost of various rainfall amounts on particular days of the year. Such an application has its dangers, however: it may be shown econometrically (Lave, 1963) that the effect of better weather forecasts for the raisin industry could be to cause the profits of the industry as a whole to fall.

Probability forecasting is becoming of increasing interest to climatologists and applied meteorologists, as is the problem of arriving at an objective decision once a reliable probability has been obtained. Our experience at Edgbaston Observatory is that many industrial and commercial users—to say nothing of farmers—have very great difficulty both in appreciating the value of these developments and in putting them into practice once they are understood. In particular, it is very rare indeed to find an organization which has financial or other records going back more than two or three years, at the most, in the necessary detail. Very careful and painstaking consultation and investigation of the past history of operations are necessary before the weather adviser can provide information for the industrialist or farmer which can demonstrably be of financial value to the latter.

In this country, there are excellent advisory services as regards weather hazards in agriculture, building and so on, which are provided free by the state and by certain university and other research organizations. Farmers, for example, may be deluged by official pamphlets and research studies advising them on how to be better farmers. My own experience is that the farmer will not take any of these publications seriously, unless he comes to know its writer personally. And, rather than wade through the mass of free (and often conflicting) information himself, he would often prefer to pay someone he knows and trusts to do the sifting for him.

My point is that under existing trends, weather information is becoming more remote and more impersonal, which the individual industrialist, business man or farmer does not like. As a canny man, he is keen to take advantage of all the free advice he can obtain from the state and from any other sources. As an intelligent man, he is suspicious of people who offer him something for nothing—or rather, he is suspicious of their methods. At present, in my experience, he is very suspicious of the paraphernalia of modern methods of mathematical decision-making—probabilities, matrices, linear programming, and so on. If weather advice is to be taken to heart, the adviser must take the trouble to work with his client and to learn the latter's problems from the grass roots. These problems are often very different from the problems that one imagined the client had, when one is in the blinkered confines of a university or research institution, or a meteorological office. For example: competition is an ever-present fear in the mind of the farmer and other businessman: of what use to him is weather information which is provided free to his competitors as well as to himself?

There are, of course, two problems involved in the economic evaluation and elimination of weather hazards. Firstly, there is the effect of weather on an industry as a whole. Secondly, there is the effect of weather on an individual operator within a particular industry. The present tendencies, towards increasing use of mathematical rigour in decision-making, towards standardization and computerization of weather forecasts, and towards centralization of meteorological services in particular regions, are all very good for an industry as a whole. However, they can be very bad for individual operators. One reason is that the latter may have greatly differing requirements as regards a meteorological basis for decision-making. Another is that what may be termed "geographical accidents" in weather situations may easily be overlooked. An example from Birmingham will illustrate this point (*Birmingham Post* and *Birmingham Evening Mail*, 1967, 1968).

Three times during the period from December 1967 to March 1968, weather events resulted in serious dislocation of economic life in Birmingham. In every case, the weather events were expected, i.e. adequate warning had been given to local authorities and other responsible bodies. But in every case, too, an "unexpected" element intervened, in the form of factors the importance of which is obvious when the local geography is considered, but the existence of which will be completely ignored if decisions are made on purely meteorological or purely economic grounds.

On the morning of Thursday, December 7th 1967, more than 80 local authorities in the Midlands were warned by Edgbaston Observatory that a snowfall of up to 3 in. was expected within the next 24 hours. Snow commenced to fall at 0200 hours GMT on Friday, December 8th, and when it ceased at 0630 hours, 3 in. had accumulated. All normal anti-snow operations were carried out in time by the Public Works Department: salt was placed on the snow as soon as it fell at about 0100 hours. However, the coincidence of a heavy snowfall between 0430 and 0500 hours with a rapid drop in temperature, and with the arrival of vehicles heralding the first rush hour, resulted in chaos. The early vehicles packed the snow severely, turning it into hard ice, at temperatures below which the salt had no effect. Gritting lorries were held up in the traffic jams, thus aggravating the road conditions, and by the end of the day, 5 million man-hours had been lost in the west Midlands, due to lateness or non-arrival of workers: this was equivalent to nearly one-half of the total time lost in strikes in one year in the British motor industry. Incorrect phasing of traffic lights at certain road junctions was one major cause of the traffic jams. Another was the fact that, since the city has more than 1000 miles of roads, only the major roads can be treated with salt and grit. Consequently, vehicles frequently became ice-bound at the junctions of minor with major roads. Even in these adverse conditions, however, some people profited: shopkeepers in outlying parts of the city experienced a day of marvellous trade as thousands of schoolchildren were sent home, and many mothers concentrated on local shops instead of going further afield.

No amount of improvement in the accuracy of weather forecasting could have prevented the traffic chaos in Birmingham on December 8th, and even if the cost/loss ratios for all the operations in the area had been known with 100 per cent precision, this would not have helped the situation. What would have gone a long way to have prevented the chaos, however, would have been a previous study of geographical influences on traffic movement in the city—for example, the delimitation of localities characterized by steep hills on minor roads close to their junctions with major roads, which, under certain weather conditions, would be expected to become ice-bound. Fore-knowledge of the location of such areas, and a broadcast warning to the public—after the general weather forecast for the area—advising them as to which of these areas were likely to be particularly dangerous after the expected snowfall, would have reduced much of the traffic chaos. My contention is that the delimitation of areas of likely traffic congestion in given weather conditions is essentially a geographical problem, and only in part a meteorological or traffic-engineering one.

On Tuesday, January 9th, 1968, rail services were crippled throughout the Midlands by what was described as the heaviest snowstorm for 17 years. In this case, snow drifting on to points outside New Street station put the new electronic signalling system out of action. 600 points occur in the Midlands area, 300 of which are key points, normally prevented from freezing by propane heaters. However, snow drifts of up to four feet swamped the heaters, with the result that the automatic signals were held at red, and New Street station was closed from 0700 to 0900 hours GMT. This was extremely unfortunate, in view of the "accident" that many commuters, on seeing the snow outside their bedroom windows, decided to leave their cars at home in order to avoid the traffic chaos of the previous month, and to rely on British Rail services. Another "accident" was the fact that British Rail relied on a national forecast, which predicted a thaw, and hence was not prepared for snow drifting, whereas Birmingham Corporation relied on a local forecast of snow, with the result that salting and gritting lorries were out in time. Most of the city's main roads were thus cleared ready for the traffic flood which did not materialize. The cost of the 4 days of adverse weather to Birmingham Corporation's Public Works Department was of the order of £55,000. This figure represents the cost incurred when local forecasts are both provided for, and acted upon, by the Department, and it is difficult to see how it could be reduced. It probably represents the basic cost of a 10-in. snowstorm to the Public Works Department of a conurbation of 2 million people, which even the most accurate weather forecasts will not be able to reduce.

On Monday, February 5th, 1968, national and local forecasts predicted snow in Birmingham during the afternoon. As a result of the forecasts (which proved to be correct), many people left work early, in order to avoid the road and rail chaos of December and January. This "accident" proved to be most unfortunate. Although the Public Works Department had adequate warning and put its lorries on the roads in time, the bringing forward of the first peak traffic period by several hours meant that the gritting lorries became caught in the jams. Thus within minutes of the expected snowfall, Birmingham became a traffic-choked city once more. British Rail services, also prepared for the snow, ran to time, but because only 3 per cent of Midland commuters normally patronize the railways, the greater proportion of the city's population did not take advantage of this.

The conclusions to draw from the events of these recent months in Birmingham is that the economic measurement of weather hazards is not merely a problem in economics and in meteorology. Perfectly accurate weather forecasts will not eliminate weather losses to the economy: they may even increase these losses, and, in any case, some members of the community stand to gain from adverse weather. Mathematically precise systems of decision-making will rarely solve the problem, because the necessary data is usually unavailable. Statistical studies of the relation between weather and economic activities are very useful in pointing to relationships which have existed in the past, but extrapolating them into the future takes no account of "accidents".

As director in Birmingham of the only British post-graduate school of applied meteorology and climatology, the writer naturally hopes that applied meteorologists may help to demonstrate how to avoid weather hazards in the future. However, unless investigations in applied meteorology, which purport to solve real problems, also take into account the local geography of the situation, and the "accidents" which could be characteristic of it, then the results of these investigations, however elegant mathematically, will be not only useless practically, but also dishonest scientifically if they are claimed to be solutions of these problems. The greater part of the expense and inconvenience suffered by the citizens of

Birmingham during the recent months could have been avoided if a realistic and sophisticated geographical analysis of the logical possibilities involved in the snow problem had been available.

REFERENCES

The figures quoted in the above Chapter are taken from accounts in the *Birmingham Post* of Dec. 9th, 1967; Jan. 9th, 10th, 11th, 12th and 13th and Feb. 6th and 7th, 1968; and the *Birmingham Evening Mail* of Dec. 8th, 1967; Jan. 9th and Feb. 6th, 1968.

BRIER, G. W. (1950) Verification of forecasts expressed in terms of probability, *Monthly Weather Rev.*, vol. 78, pp. 1–3.

EPSTEIN, E. S. and MURPHY, A. H. (1965) A note on the attributes of probabilistic predictions and the probability score, *J. Appl. Met.*, vol. 4, pp. 297–9.

GRINGORTEN, I. J. (1959) Methods of objective weather forecasting, *Advances in Geophysics*, vol. 2, pp. 57–85.

KOLB, L. L. and RAPP, R. R. (1962) The utility of weather forecasts to the raisin industry, *J. Appl. Met.*, vol. 1, pp. 8–12.

LAVE, L. B. (1963) Investigation of the value of better weather information to the raisin industry. *Econometrica*, Vol. 31, pp. 151–64.

MURPHY, A. H. (1966) A note on the utility of probabilistic predictions and the probability score in the cost/loss ratio decision situation. *J. Appl. Met.*, vol. 5, pp. 534–7.

SHORR, B. (1966) The cost/loss utility ratio. *J. Appl. Met.*, vol. 5, pp. 801–3.

THOMPSON, J. C. and BRIER, G. W. (1955) The economic utility of weather forecasts. *Monthly Weather Rev.*, vol. 83, pp. 249–54.

CHAPTER 9

Edited Report of the Discussions Held at Symposium XI, 1968

JAMES A. TAYLOR

I. THE DISCUSSION

Mr. Stansfield† expressed the view that contract farming in general streamlined decision-making for the farmer or grower, promoted greater efficiency and often afforded better protection against particular weather hazards. Decisions were made more promptly, e.g. mechanized harvesting, and could combat a weather hazard which may cause severe losses to the individual farmer dependent solely on his own equipment. Mr. Taylor pointed out, however, that contractors in general had to cater for the "average" season or, optimistically, one somewhat above the average. Good harvests and over-production could be as embarrassing as bad harvests and under-production. The variable pea harvests in south-west Lancashire were a case in point. The economic disposal of unharvested pea crops in glut years, when the field services and the canning and freezing factories of the contractors were unable to cope with the supply, was just as formidable a problem as the low cash returns payable for a poor quality pea-crop in a bad season, be it due to unsuitable weather or inadequate management. It was difficult to reconcile efficient contracting at all times with efficient husbandry.

Mr. Taylor reported on the work of Chambers (1964), which provisionally attributed 25 per cent of the variation in milk yields in the Wiltshire–Dorset–Somerset district to weather factors, and asked Mr. Stansfield whether, in the light of his experience, he could offer any parallel estimate of the impact of weather factors on farm management decisions and farm

† An alphabetical list of discussants is given at the end of this chapter on page 160.

practice, as related to consequentvariations in productivity. Mr. Stansfield replied that although there would be much variation from farm to farm, the range would be of the order of 15 to 30 per cent depending on the enterprise concerned. Values of the order of 25 per cent would be quite characteristic.

Mr. Hogg, in reply to a question on the costing of alternative methods of frost protection, stated that some years ago the National Institute of Agricultural Engineering studied all available methods of frost-protection. The use of wind-machines was shown to be technically feasible but uneconomic in practice. Mr. Wardle stressed the need to examine low, as well as high, probabilities of frost risk. For example, what strategy would cater for the rare, catastrophic frost of every 30 or 40 years? How could the cost of frost protection devices be reconciled with the infrequency of the hazard? Mr. Hogg intimated that sites or enterprises liable to one major killing frost over 30 or 40 years would probably be more frost-liable in the interim than was often fully realized. This was his experience for example in certain districts of Somerset. Variations in crop yields due to minor frosts could be very substantial. The question to be answered was, could frost protection be provided at less cost for once in 30 years than for three years in 10? Mr. Taylor commented that mossland farmers in Lancashire to date accepted a main crop potato failure once every 10 to 20 years in view of the satisfactory profits of all the intervening years, taken together. They had not used frost-protection devices to date, except for one or two individual farmers near Rainford who have successfully used simple heaters and smoke devices when severe, late radiation frosts were imminent. Mr. Hogg in reply to a question from Mr. Toulson about sources of information on irrigation-need, referred to his recently published irrigation atlas of England and Wales (Hogg, 1967).

Dr. Gloyne raised the subject of the application of statistical assessments of wind data which in themselves were now fairly well understood. The question remained, however, as to what characteristic or critical wind caused tree-damage or tree-collapse. Was it mean wind, a gust of x seconds, or a sharp change of wind speed or direction? Was not wind-throw probably associated with a pre-conditioning of the soil? Mr. Wardle replied that the soil categories used in his paper were very broad and varied very little seasonally from the engineering point of view. On the other hand drainage could modify the soil locally and affect the susceptibility to wind-throw of trees. Topography too was an important local variable, often causing unexpected changes in exposure or shelter. Our understanding of the relationship between air flow and the geometry of the underlying surface was being gradually improved.

Mr. Smith and Mr. Wardle referred to the successful liaison which had been established between the Meteorological Office and the Forestry Commission in the anticipation of fire risk. From day to day the weather situations were directly and locally evaluated so as to implement economically and effectively all available fire protection procedures.

Mr. Joce expressed surprise that the peak danger period for fire in forests came so early in the year and not towards the end of the summer. Again, the data for the extremely dry year of 1959 gave well over 5000 outbreaks but a very low acreage of damage. Did this imply that at times of maximum risk, maximum control on outbreaks was exercised keeping damage to a minimum? In contrast, at times of negligible risk, only a few outbreaks could cause much damage by wide and rapid spread of the blaze because the protection services were not fully on the alert.

Mr. Wardle concurred in general but pointed out that even if all the Commission's 500,000 acres of forests in Scotland, for instance, were placed under observation by all available fire protection services, it would still be possible for a major outbreak of fire to be entirely missed.

Mr. Taggart noted that two dry years, 1955 and 1959, were followed by two years of high acreages of damage to forests by fire, viz. 1956 and 1960.

Mr. Smith and Mr. Taggart referred to the effects of a dry winter in increasing fire risk in the subsequent spring, especially when the surface grasses, especially Molinia, dry out—this was the most "inflammable" of all situations.

Mr. Hogg asked whether wind-blow was more often than not caused by strong winds from a non-prevailing direction. Mr. Wardle was of the opinion that most wind-blow was due to prevalent winds being stronger than usual. He also stressed the rapid diminution of tree-stability with tree heights of more than 30 ft and especially above 50 ft.

Mr. Smith inquired about the sequent effects of wind-blow. Did trees collapse individually, exposing adjacent trees? Or did trees blow at random at all stages? Again, could a plantation be streamlined to steer wind flow over it and reduce or even avoid damage. Mr. Wardle explained that the pattern of wind-throw depended, among other factors, on the wind-speed. When the latter was at or about the critical threshold, then blow was individual and random. Above the threshold, entire areas of forest were liable to blow. The "landscaping" of forests to wind flow was possible through both planting and thinning programmes, subject to the satisfaction of economic arguments.

Mr. Smith defended the conventional types of synoptic weather forecast available in Britain but Dr. Stringer felt they could be extended and more locally and specifically applied. He emphasized the circumstantial complications which could exacerbate an imminent weather hazard. Mr Hogg interjected that it was easy to be wise after the event but Dr. Stringer insisted that a more sensitive and "applied" forecasting system could ease many problems of production and communication and enable major economies to be achieved. Mr. Wardle indicated, however, that in actual practice, the problem of using weather forecasts was essentially educational. People who make decisions on the basis of weather information should be fully conversant with such information and its local or specialized implications, and should liaise adequately with cognate services or operations.

Dr. Gloyne referred to the pattern of potato blight incidence in the British Isles. In Ireland spraying was more or less automatic since the risk of disease was always present. In England, attempts were made to identify areas where the disease was most likely to occur, so spraying is intermittent and, in a sense, optional. In between these two extremes is a point where the change-over occurs from decisionless routine to compulsory decision-making. This point was crucial to all these discussions—i.e. the point at which decision-making machinery comes into play, and later the extent to which it should be allowed to come into play. Mr. Taylor pointed out that wholesale accurate implementation of forecasts could flood and ruin a market. For example, if every Pembrokeshire grower of early potatoes applied irrigation successfully and accurately without notice the effects on prices and markets could be far-reaching.

Mr. Joce, recounting his experiences as farm management adviser, stressed the inevitable role of the management factor in the application of specialized information to real situations. Improbable or chance factors in addition to the probable and predictable factors had always to be taken into account.

Mr. Toulson, commenting on Mr. Tyrrell's paper, said that the U.K. production of main-crop potatoes was already falling and this could well happen also to "early" production. A contraction of the acreage had already occurred in Cornwall and, if this occurred in Pembrokeshire, Tyrrell's paper would be useful in classifying early-potato land and calibrating its possible contraction.

Dr. Gloyne, commenting on Mr. Taggart's paper, stressed the rapidity of the changes in climatic conditions which occur in Scotland with increases in elevation and/or in distances from the sea. These affected farming systems and vegetation markedly. Yet, the contrasts were scarcely visible in the available conventional meteorological data.

Mr. Wardle, commenting on Mr. Taggart's paper, designated the holdings adopted in the paper as "computer farms". Had Mr. Taggart considered treating the holdings as one single farm? If so, extrapolations for forestry purposes might be feasible as regards the optimal deployment of a labour force. Mr. Taggart reported on his extensive work on this subject. The combination of labour forces invariably led to economies. Mr. Brooke pointed out that Mr. Taggart, in measuring the effect of weather on labour availability, had presupposed that he was able to measure the effect of the weather on the availability of work-days. Mr. Taggart accepted that this was the case, and that it conditioned his findings.

Mr. Rubra, opening the general discussion, referred to the use of the term "cost/benefit analysis". Much of the discussion in the Symposium, he felt, concerned costings rather than cost/benefits. In ordinary accountancy, the object was to determine how cheaply a particular objective could be achieved within the frame of one working unit. Cost/benefit analysis usually tried to widen this to include "social" costs, i.e. costs that did not acrue directly to the person who was making the expenditure. If people were delayed on their way to work by the consequences of bad weather, this was a "social" rather than a "financial" question. The distinction did become a little blurred in the case of the Forestry Commission which was in business on the one hand as a commercial enterprise but which, on the national scale, received subsidy and provided amenity.

Dr. Stringer warned against concentration on the costs rather than the benefits in the equation. He also differentiated between individual and national costs and benefits. Quite often the cost/benefit of the individual operator was completely different from that of the next one. Mr. Rubra pointed out that individual farmers may make entirely opposite uses of the same weather information on the same occasion. All that more specific weather forecasts, etc., for farmers meant was that more accurate information was being fed into the economic models, viz. the farms. In addition, information was required on the extent to which farmers were *using* this information, before accurate assessments of factors involved could be made.

Mr. Wardle expressed the hope that he did not give the impression that the Forestry Commission were able to make decisions on fire protection and wind-throw protection on the basis of costs alone because the benefits from avoidance of losses were an equally significant part of the equation. He added that he had approached the problems specifically from the point of view of the commercial operator concerned with his investment problem. However, severe cases of wind-throw in Britain had led to the introduction of social costs. In the 1953 gale it was regarded as appropriate that the government should provide a special subsidy to mobilize the timber and transport it to the mills before it deteriorated. In Germany, timber blown in gales was regarded as not taxable. Costs were thus shared between private and public sources which, in this instance, appeared equitable.

Mr. Taylor initiated discussion on the subject of weather insurance as a possible mirror image of the cost of weather hazards, and asked Dr. Stringer if he knew of the extent of weather insurance in his service catchment in the Midlands.

Dr. Stringer replied that weather insurance was increasing at a very great rate among farmers, builders and civil engineers. The insurance companies were interested in objective decision-making on meteorological grounds. The trouble was the lack of data. Some companies were quite interested in theoretical meteorology as a means of estimating the weather probabilities and insurance rates. Dr. Stringer estimated that all fruit growers took out insurance and 50 per cent to 60 per cent of market gardeners, certainly the larger ones. Last year many blackcurrant growers lost much of the crop, but they were not particularly worried since they were insured. Some growers, however, were investigating the possibility of setting up wind-machines and irrigation protection but in actual fact insurance would be cheaper.

Mr. Smith expressed the view that financial insurance was an improper use of resources. We ought to be looking at physical ways of weather-proofing rather than financial weather-proofing. Mr. Wardle regarded insurance as a perfectly legitimate type of investment. Mr. Rubra remarked that insurance was legitimate from the point of view of the individual, but, being a transfer payment, no net social benefit accrued.

Mr. Taylor then switched the discussion to subsidies. In Britain specific subsidies were available for specific types of agricultural practice or improvement e.g. for hill farms, ploughing of upland pastures, improvement of approved farm buildings and, more recently, deficiency payments, e.g. for barley. Many of these subsidies applied to areas which may be described as climatically inferior or deficient. As applied agro-meteorology advanced and intensification of methods exposed more farmers to the mercies of unavoidable weather hazards or the costs of protection against such hazards, would the future witness an extension of subsidies to meet these contingencies—as an aspect of a late twentieth century policy to preserve a shrinking countryside?

Mr. Taggart thought this implied prediction was true. In Scotland the general argument that the lowland farmer required special subsidy as well as the upland farmer, in special circumstances, was accepted.

Mr. Taylor, continuing his thesis, stated that the large producer was currently immune to the argument since he could, for example, carry one or two bad years in ten and survive, making a quite substantial profit over 20 or 30 years. The small producer, in contrast, with the smaller turn-over, was more vulnerable to the occasional bad year and could be rendered bankrupt without subsidy when the extreme hazard occurred. Was the small grower worthy of subsidy because of his frost hazard every 5 years or his irrigation-need every six years in 10, for example, just like the hill farmer was on the grounds of his annually persistent hill climate?

Another point of view expressed was that an elimination of *all* existing special subsidies would be preferable. The subsidies dated back to times of national emergency, to years of depression and wartime but had now become accepted. They were social, rather than financial, measures to help maintain so far as possible the socio-economic structure of rural areas. But all ethical arguments aside, would it be wise to try to extend an artificial situation into wider sphere? Mr. Rubra considered that this all hung on the precise object of subsidy policy in the first place. It could be oriented to prevent hardship on individual holdings, or conversely, it could be part of a regional strategy to arrest socio-economic decline in rural areas.

Miss Morgan and Mr. Griffiths pointed out that climatic considerations only indirectly confirmed the payment of subsidies to hill farmers. Lowland farmers produced crops for example which attracted guaranteed prices. Upland farmers were often concerned with intermediate products, e.g. young animals, for which prices might well be subject to more variation. Again, a farmer growing 500 acres of barley was more subsidized than a farmer growing 10 acres. Similarly, a man with 1000 ewes was more subsidized than a man with 100 ewes. These policies could hardly be said to be perpetuating the smaller farms.

Mr. Wardle, referring to Mr. Hogg's paper, wondered whether the relationship between frost probabilities and investment costs could be studied as a variable of the investment cost. He (Mr. Hogg) did in fact mention there were different types of probabilities that could be adopted as a standard one, namely the succession of 5 or 6 days of frost and the investment in his million-gallons water tank for irrigating a crop of blackcurrants. What interested him (Mr. Wardle) was that he (Mr. Hogg) could seriously consider the very substantial investment of a million gallons of water for a mere 10 acres of blackcurrants. Would there not be sharply diminishing returns from the investment as the tank was increased from, say, 100,000 to 200,000 to 300,000 gallons and so on? The Forestry Commission had explored this with reference to degrees of wind protection and the effects of different intensities of drainage.

Mr. Hogg admitted that the figures given, were capable of greater refinement but stated his task as a meteorologist was to demonstrate this special use of meteorological data. It was up to the N.A.A.S. to work applications of different levels of the effects of protective irrigation.

Dr. Edwards welcomed the increasing information which was emerging on weather probabilities. Every farmer worked on the basis of probabilities. He thought that every arable farmer was over-mechanized to an unknown degree. This was costly over-investment against the possibility of very infrequent and often improbable losses.

Mr. Rubra cited a simplified, imaginary example. Suppose a weather hazard, if it occurred, would cause losses valued at £3000 and the average expectation that it would occur is a one in twenty chance; then it follows that 1/20th of £3000, or £150 per annum, is the maximum to spend on preventive measures. If it was neccessary to invest a larger capital sum in one year, as is likely for protective measures as opposed to insurance, then "discounted cash flow" would have to be used as though the money were borrowed on a mortgage and repaid annually. This means that the maximum capital works which could be justified would be much less than £3000 since the choice is either to take preventive measures *now* or possibly (or certainly) losing a crop at some unspecified time in the future. On the probability here stated (5 per cent, or one disaster in 20 years) the best average assumption is that it will occur 10 years ahead, so interest costs should be spread over this period. Deducting these at 7 per cent compound interest over 10 years gives a maximum expenditure of almost exactly half the expected risk, i.e. £1518 7*s*.

Mr. Wardle stressed the need also to cater for the moderate or average hazard which occurred more frequently and could cause substantial, cumulative loss.

Mr. Clarke differentiated between the precautions needed to maintain average profitability levels and those needed to cope with the unlikely catastrophic event.

Mr. Hogg concluded by saying that all the assumptions used in the discussions had been based on probabilities and incorporated the idea that the British climate itself was not changing. Recent work by Lamb (1966) for instance had demonstrated the nature and scale of recent and current climatic trends in Britain, trends which may well have effects on agricultural practices and productivities.

II. IMPLICATIONS AND CONCLUSIONS

1. *The Entrepreneur*

The essential focus of these papers and discussions of the impact of weather hazards on economic systems is the entrepreneur himself whose varying personal calibre and technical capacity will cause differences in productivity from place to place and from season to season. At the same time, selected materials or processes used in production, not least the almost ubiquitous factor of transportation, will be more or less susceptible to weather hazards which, in their turn, will indirectly affect the standard and range of productivity. The decision-making of the entrepreneur is based partly on a personal prediction of how operative factors, individually and collectively, are going to impinge on marketing conditions and ultimately, profits. In other words various patterns of probability have to be estimated. Such estimations may be based on the cumulative experience, short-term or long-term, of the individual entrepreneur or groups of entreprenuers. The experience of previous generations of entrepreneurs is inherited and may pre-condition the estimation of probabilities. This applies in particular to farming communities especially where enterprises are specially adapted to regional advantages of climate of soil. Again, however, the estimation of such probabilities as those of the seasonal weather is perpetually subject to amendment in the light of new expansion and of any new external stimuli which may bring innovations and new levels of production.

The extent to which the entrepreneur plans ahead to accommodate weather hazards depends on the scale of his operations, and the degree of intensity and specialization of his production. In general, the larger the operation or the more intensive the system, the more frequent and more sophisticated the weather-proofing becomes. This can apply both to large units, which may achieve only modest levels of production (e.g. certain large hill farms), or to small units which are geared to very high levels of production (e.g. certain market gardens). Again, the larger unit or the intensive unit are more able to cater for, or survive, the extreme hazard as distinct from the average or expected hazard. Protection could be achieved by the purchase and mounting of physical devices (e.g. irrigation against drought or frost) or by the establishment of proper and specific financial insurance against losses. Survival is often based on the immunity provided in a disaster year by the cumulative surplus profits which accrue in the more frequent average and good years. In contrast, a small under-producing unit often lacks the resilience to improve production and protection in order to combat the extreme hazard. Vulnerability to weather hazards thus varies on a wide scale according to many factors associated, directly or indirectly, with the entrepreneurs involved.

2. *The Variability of British Weather Hazards*

Some of the major hazards affecting Britain, and British agriculture in particular, include gales, floods, prolonged wet spells, frosts, droughts, and storms (e.g. of rain, hail or lightning), and special circumstances when weather vectors bring or spread disease, industrial pollution or sea-salt. The average "equable" image of the British climate disguises the wide and quite extreme range of British weather which because of general unpredictability and wide variation with altitude, is difficult to forecast *locally* or even regionally. Again, weather hazards often involve combinations of weather elements and although particular areas of Britian and particular times of year are characterized by the occurrence and intensity of certain weather hazards, the exceptions to the rule occur in themselves too frequently for comfort, usually because preparations for the unexpected are bound to be inadequate—e.g. the floods in mid-September 1968 in south-east England.

Examples are legion of the way combinations of weather elements constitute the major weather hazards. Gale-force winds may deplete ripening cereals and fell trees but they are much more damaging if they are charged with cold air, which, if moist, will cause intense exposure. Strong, drying winds can be equally harmful. Gale-force winds plus snow produces blizzards and drifts which can lead to heavy stock losses in the uplands. Again, strong winds off the sea spread sea-salt well inland but in industrial areas strong winds can disperse industrial pollution. Heavy, intense or prolonged rain all produce flooding but if the ground is already saturated before the rain begins, or if the rain is coincident with a high tide and onshore winds in a coastal district, then the flooding is severely increased (e.g. the East Coast floods of 1953, and see also Rutter and Taylor, 1968). High level or sloping sites are not necessarily immune to flooding. The effects of flash floods after storm rainfall in mountainous or hilly areas is to cause severe local damage and, for example, soil erosion and landslips, often at new, as well as old, points of drainage convergence.

Frosts are winter phenomena but the critical frosts are those of late spring and early autumn which determine the length of the growing season. Basin sites suffer extremely from radiation frosts; wind frosts affect slopes oriented against the direction of wind flow; in the uplands of Britain frost frequencies are high at all sites regardless of topography. On the other hand, at sites where the relief is flat and constant, sandland will always be earlier than clayland but suffer from later frosts, whilst peatland will be the latest of all three and suffer the latest frosts, *ceteris paribus* (see Taylor, 1967).

Whilst particular sites or districts are notoriously susceptible to particular weather hazards at particular times of the year in most years, much of the British Isles is liable at any time to receive the effects, directly or indirectly, of any of the common weather hazards. Nonetheless, some areas are more liable to specific hazards than others. It is currently impossible to rationalize the relative frequencies of these hazards in a manner amenable to the weather-proofing needs of individual entrepreneurs. The type of computer that would be needed to assess the probabilities in both time and space to the required degree of refinement has not yet been made available to the Meteorological Office. None the less, by the year 1972 and by the use of a computer costing £3,000,000, the Office anticipates that forecasts of the *amounts* of rainfall should be possible (Meteorological Office, Annual Report for 1967). More reliable five-to-seven-day forecasts are also envisaged.

Even so, it is the individual entreprenuers on individual holdings who receive the effects of the individual, unpredictable hailstorm or of the anomalous killing frost or severe flood. It is unlikely that more refined weather forecasts, no matter how punctual, will ever come to the rescue of the type of individual producer who either by lack of capital or know-how or inclination, will not invest in physical or financial weather-proofing. Unfortunately, this type of producer is all too numerous especially with reference to small and medium-sized farms. This widespread vulnerability to weather hazards is in itself an impediment to intensification of production. Thus, a vicious circle is created.

3. *Weather-proofing*

Simple avoidance of weather sensitive sites or enterprises is a traditional British farmer's way of weather-proofing. However, the inducements offered by changing markets and prices and by specific government legislation, have tended to push the rotation of farm enterprises more indiscriminately and more frequently around the farm. Again, the recent conversion of such a wide range of holdings to liquid milk production in Britain has led to an almost permanent acceptance by many of the variations in production, often substantial, due to the weather. There are many thousands of small dairy farms in Britain, often below 50 acres in size, which find it increasingly difficult to intensify or even to continue to make ends meet. Extra-farm income has been the solution in many areas. These small dairy farmers often lack the capital and/or inclination to buy physical or financial protection against weather hazards. Duckham (1964) has demonstrated admirably how the larger, or the enterprising smaller, dairy farmer in south-east England can choose between (a) weather-proofing against drought and maximizing net profits for most years, or (b) not taking any steps to weather-proof, and underproduce except in the wet years and risk heavy losses in the drought years.

The wealthy farmers and growers, regardless of size of holding, are able to invest in weather-proofing. Similarly, the big cities may possibly be able to afford types of protection against, for example, winter weather hazards. Small towns in contrast are less able to do this and they are far more numerous. As Stringer has shown herein (Chapter 8, pp. 88–91), even the best laid plans to combat the weather can fail to succeed because of bad liaison between all the parties concerned. Such circumstantial factors as created the chaos in Birmingham last winter (1967/68) are a source of further variation—often quite unpredictable—in the potential vulnerability of sites or systems to weather hazards. The final argument, however, must turn on costs, on profits, on losses and on cost/benefit analysis.

4. *The Economic Factor*

Much of the decision-making in the sphere of weather-economics must consider a variety of costings and several arguments affecting cost/benefits. As is shown in some of the foregoing studies the measurement of these economic factors is often difficult and in some instances virtually impossible. Amenity benefits and national or regional benefits—and costs—are elusive of definition. It is essential to differentiate between economic considerations affecting (a) the entrepreneur, (b) the region, (c) the nation. A *laissez-faire* economic system offers further difficulties in achieving these differentiations. However, in recent decades the degree of interference with the traditional *laissez-faire* system has been increasing. Much direct legislation via marketing and subsidies for example has come to bear on agriculture, and forestry itself in Britian is largely a government sponsored enterprise which is in a peculiar category since it is expected in the long-term to achieve some degree of commercial viability. The crucial issue, however, is the relationship between these public, and the several private, forms of financial inducements and the response of the individual farmer or grower at a time of weather stress which may, or may not, have been anticipated by either side. A given set of relationships may become fairly standardized over time. Actuarial assessments are based on historical inventories of previous cases. However, if the hazard increases (or decreases) in frequency (due perhaps to a short-term climatic trend) or if the level of premiums has to rise (due for instance to excessive claims being made), or if the premiums are reduced (due to excess profits) or if subsidies are changed by legislation—all these individual changes can alter the entire system.

Weather-proofing in Britain until recently has been limited and *ad hoc*. There are possibility signs of change and of a more serious approach in several areas. There is certainly room for a more accurate costing of weather hazards, but even if this were achieved, the vexed question would remain as to how advice on weather economics could be effectively and equitably disseminated without upsetting the competitive balance of existing agricultural and other economic systems. The final reconciliations probably lie in the philosophy that a fundamental basis of planning is the areal adjustment of investment from all sources public and private to attain economic optimization of resources on both a national and regional basis. A corollary of this is that the net cost of easing or controlling constraints in a producing system should be adjusted to the areal and temporal probabilities of the instrument or hazard concerned. In other words, a case could be made, if necessary, for the establishment of a permanent national disaster fund as a national and regional protection against weather hazards. This would help to replace the *ad hoc* and generally inadequate system of compensation currently practised. At the same time, steps should be taken to prevent opportunist exploitation of such a fund either in anticipation of the hazard or subsequent to its occurrence.

REFERENCES

CHAMBERS, RUTH (1964) Weather hazards and milk production: a review, in *Memorandum No. 7, Major Weather Hazards affecting British Agriculture*, Ed. J. A. Taylor, Geography Department, University College of Wales, Aberystwyth.

DUCKHAM, A. N. (1964) Weather and farm management decisions, in *Memorandum No. 7, Major Weather Hazards affecting British Agriculture*, Ed. J. A. Taylor, Geography Department, University College of Wales, Aberystwyth.

HOGG, W. H. (1967) *Atlas of Long-term Irrigation Needs for England and Wales.* M.A.F.F., London.

LAMB, H. H. (1966) *The Changing Climate.* Methuen, London.

METEOROLOGICAL OFFICE (1968) Annual Report for 1967, H.M.S.O.

RUTTER, N. and TAYLOR, J. A. (1968) Intense rainfall in West Wales, *Weather*, vol. XXIII, No. 3, pp. 94–100.

TAYLOR, Mr. J. A. (Ed. with contributions) (1967), *Weather and Agriculture*, pp. 15–36, 37–47, 231–25. Pergamon, Oxford.

LIST OF DISCUSSANTS

BROOKE, Mr. M. D. Farm Management Adviser, N.A.A.S. (West Midlands Region), M.A.F.F., Woodthorne, Wolverhampton, Staffs.

CLARKE, Mr. J. A. Lecturer, Department of Physiology and Environmental Studies, School of Agriculture, University of Nottingham, Sutton Bonington, Loughborough, Leics.

EDWARDS, Dr. R. S. Senior Lecturer, Dept. of Agriculture (Crop Husbandry), University College of Wales, Aberystwyth, Cards.

GLOYNE, Dr. R. W. Senior Meteorological Officer, Meteorological Office, 26 Palmerston Place, Edinburgh 12, Scotland.

GRIFFITHS, Mr. D. J. Senior Research Officer, M.A.F.F., A.L.S., Great Westminster House, London, S.W.1.

HOGG, Mr. W. H. Senior Meteorological Officer, N.A.A.S., Government Buildings, Burghill Rd., Westbury-on-Trym, Bristol.

JOCE, Mr. E. H. B. Farm Management Officer, N.A.A.S., Block 2, Government Buildings, Lawnswood, Leeds 16, Yorkshire.

MORGAN, Miss J. P. Senior Research Officer, M.A.F.F., A.L.S., The Crown Building, Plascrug, Aberystwyth, Cards.

RUBRA, Mr. G. N. Lecturer, Department of Economics, University College of Wales, Aberystwyth, Cards.

SMITH, Mr. C. V. Senior Meteorological Officer, M.A.F.F., Block C, Government Buildings, Brooklands Avenue, Cambridge.

STANSFIELD, Mr. J. M. Deputy Farms' Director, University of Reading, University Farm, Sonning-on-Thames, Reading, Berks.

STRINGER, Dr. E. T. Senior Lecturer, Department of Geography, The University, Birmingham.

TAGGART, Mr. W. J. Lecturer, The Edinburgh School of Agriculture, Department of Agriculture, University of Edinburgh, West Mains Rd., Edinburgh, Scotland.

TAYLOR, Mr. J. A. Senior Lecturer, Department of Geography, University College of Wales, Aberystwyth, Cards.

TOULSON, Mr. G. A. Crop Husbandry Adviser, N.A.A.S. (Wales), Trawscoed, Cards.

WARDLE, Mr. P. A. Planning and Economics Branch, Forestry Commission Research Station, Alice Holt Lodge, Wrecclesham, Farnham, Surrey.

CHAPTER 10

Economic Postscript

G. N. RUBRA

IT WILL already be clear that weather has a direct influence on a wide range of economic activities. Ordinary weather conditions such as rain and sunshine affect predominantly outdoor occupations like farming and tourism, whereas manufacturing industry is normally affected only by exceptionally severe conditions (fog, flood, heavy snowfall, etc.) because these delay, or in some cases prevent, workers from reaching the factories. The private householder, farmer, and forester also suffer from some of these severe conditions since frost, gale, and flood can damage buildings, crops, standing timber, and even livestock. Road transport falls somewhere between the two groups as an outdoor industry with limited weather protection. Severely affected by fog, snow and flood, traffic can also be delayed by ice, high winds and heavy rain. Since it is necessary to drive more slowly, journey times are lengthened and congestion is increased because more vehicles are on the road at any one time. Even so, there is some evidence that a proportion of drivers make insufficient allowance for bad weather conditions resulting possibly in more accidents per vehicle-mile than in fair weather. Unfortunately, no detailed work on this particular point was presented during this symposium.

What is clear, however, is that the possibility of taking protective measures has altered the nature of weather as an economic input. Whereas it was looked upon in the past as a capricious event which could not be influenced at any price—one simply had to live with it—it can now be seen that money spent on precautions can convert the absence of adverse weather effects into a saleable commodity and hence a factor of production. Putting radiators and artificial light into a hothouse or henhouse are other ways of paying for (simulated) weather effects through the use of ordinary economic resources. Although it is not at present profitable to use such a controlled environment for, say, wheat growing, such concepts are leaving the era of the pipe dream to become mere exercises in costing.

To clarify the specifically economic issues, small economic units like farms need to be considered separately from large ones such as public corporations. The determination of the limits to precautionary expenditure, a problem which all units of organization have in common, will be dealt with last in this essay.

The farmer has four basic courses of action open to him in response to weather hazards, apart from actual farm-management techniques such as rotation and timing:

(1) avoid weather-sensitive enterprises as far as possible;
(2) ignore possible losses from adverse weather;
(3) insure against possible losses from adverse weather;
(4) install physical protection against adverse weather.

A rational choice among these should be dictated by balancing expenditure against expected yields with some discounting for uncertainty. In other words the farmer's aim will be to maximize his expected private net profit, though he may choose to settle for a safer but more modest return rather than gamble on a riskier, but potentially more profitable, enterprise. Both McFarquhar (1961b) and Camm (1963) showed that the variation in expectations from riskier enterprises in 95 per cent of cases is not so great as the mean improvement which these offer over the safer enterprises, so the incentive to switch to safe crops is small (scheme 1 in the table in Fig. 46 (p 110). Their work assumes that output and timing have no effect on prices, which would not hold if the conditions experienced were widespread; but as the alternative assumption would tend to *decrease* the variability of receipts, the result is not affected. Langley (1966) has also published some work on variability of farm incomes, which he ascribes partly to weather effects. He considers percentage changes in year-to-year incomes (rather than profits) and assumes an absence of trend (over 10 years). The figures are corrected for price changes. Effects resulting from management policy are not excluded. Even so, the variation in the total imponderables facing the farmer, including as it does some effects which he would be prepared for, is relatively less than in transport and textiles in the same period (1956–66) and, until 1965, in chemicals, though a little greater than in the engineering, electrical, and retail groups. The amount of variation was more, but the proportion less, for the larger and for the more profitable farms (two distinct, but overlapping, groups). These units, larger in terms of disposable capital, were better able to allay unforeseeable fluctuations in income though not, apparently, by means of diversification. Langley, however, was not entirely satisfied that the diversification of enterprises in his sample was representative.

As to insurance, Dunford (1961), interrogating a small sample of 41 farmers, did not find any who insured their crops against bad weather, though a few were covered for animal diseases, all for fire, and three-quarters held life policies. There is thus little evidence that either management strategy or choice of enterprise is dominated by consideration of weather hazards. Although schemes (1) and (3) have been retained in the analysis for the sake of completeness, it seems likely that in practice the choice will usually be between schemes (2) and (4).

There has also been reference, in the foregoing discussions, to cost/benefit analysis. The object of this type of analysis is to apply welfare criteria and consider not merely the profit to the individual farmer, but the balance of advantages and disadvantages to the community as a whole, insofar as these can be ascertained. It is clear from the table (Fig. 46) that the maximum benefit to the community will not automatically coincide with maximum profit for the individual farmer. This depends partly on whether he is affected by local or general weather conditions but, taking universally favourable conditions as the norm, the following patterns of relative divergence could be expected. Clearly this is grossly oversimplified and the qualitative outcomes stated serve only as an illustration. Quantification in money terms could be achieved at least in principle but, even as it stands, it shows why insurance, though saving the luckless farmer (case 3b), helps nobody else. All that happens is the spreading of risks over a larger number of farms or a longer period of time. Conversely, protection of crops (and by analogy livestock or any asset) from the weather, though more expensive at the outset, does help the community if part or all of the harvest is saved and prices kept down (scheme 4). High prices help the farmer only at the expense of the public, who can benefit from the magnitude of the physical yields but not directly from the value of the farmer's income.

Economists as a profession are unenthusiastic about subsidies, although some form of agricultural support is fairly widely accepted on strategic grounds. However, if farm subsidies are to be justified at all, strategy must be a subservient part of a wider aim, viz. that the farmer's personal interests should thereby be brought into line with those of the community. In a

Scheme chosen by farmer	Sub-grouping of schemes	Weather conditions f: favourable u: unfavourable (to chosen enterprises)		Private outcome to farmer	Welfare outcome to community
		local	national		
(1) Safe crops	a	f	u	Adequate. Chance of windfall profit excluded by avoiding risky enterprise.	Acute shortage of risky commodities, with high prices. Adequate supply of safe goods.
	b	u	f	Poor but not disastrous. Moderate crop, low prices.	Slight price increase for risky goods.
	c	u	u	Some rise in prices to offset reduced yield.	Shortage of risky crops severe with very high prices. Slight price rise for safer crops. Imports may strain balance of payments.
(2) Risky crops without protection	a	f	u	Successful. Good yield and high prices.	Shortages and high prices, but more evenly spread than in 1a.
	b	u	f	Disastrous. Poor yield and low prices.	Hardly any increase from normal prices.
	c	u	u	Poor yield but high price.	Considerable shortage of risky crops with decidedly high prices, though less extreme than 1c. Some rise in safe commodities also. Imports needed marginally less than 1c.

Fig. 46. Comparison of community benefits with farm profits for various approaches to weather risk.

FIG. 46. (Continued)

Scheme chosen by farmer	Sub-grouping of schemes	Weather conditions f: favourable u: unfavourable (to chosen enterprises)		Private outcome to farmer	Welfare outcome to community
		local	national		
(3) Risky crops insured	a	f	u	Successful, but cost of premium to deduct.	As 2a.
	b	u	f	Compensated by underwriter.	As 2b.
	c	u	u	Poor yield, high prices supplemented probably by some compensation.	As 2c.
(4) Risky crops protected e.g. by watering, wind shields, drainage	a	f	u	Satisfactory. Cost of protection to deduct. Price rises will be limited if other farmers also protect.	Price rises can be controlled if crops are saved, though the cost of protection will cause a smaller increase (spread over several years).
	b	u	f	Low prices; yield maintained.	No rise in prices.
	c	u	u	Crop saved at somewhat enhanced prices; less cost of protection.	As 4a.

sense the support prices for cereals attempt this, since they protect farmers from being paid less *in toto* for a bumper harvest than for a meagre one; but in any case individual farms do not account for a large enough proportion of the total crop to enable them to maintain prices by restricting their output. On the contrary, the lower the market price falls, the more scrupulously the farmer must maximize his output to maintain his income.

Whatever the position in industry, farming is still subject to the constraints of perfect competition. Without entering into the vexed question of whether subsidies serve to optimize, or prevent the optimization of, the mix of farm enterprises in Britain as a whole in terms of net yield per acre, it is clear that it would benefit the community to encourage farmers to use what measures of physical protection are available, but not to insure against financial losses. Grants from public funds, or low-interest loans, or both, might therefore be considered for the former type of investment. (It can be argued that there may be good reason to encourage insurance too, depending on how insurance firms invest the premium income in their possession: it is conceivable that this might be in some enterprise which yielded a more productive rate of return on capital then crop protection. However, it seems fair to say that, if the community wishes to sponsor investments which insurance firms might have supplied given sufficient business, it should do so directly rather than through the medium of the farm support system.)

Two other points from previous chapters have a bearing on the relationship of the private profits of farmers to community welfare. One concerns early potatoes (Tyrrell, Chapter 6, pp. 54–5). It will pay a farmer to lift a reduced crop while the price is high rather than leave it to mature and secure say 25 per cent more yield at 30 per cent lower price. Assuming the land is not needed for urgent re-use, the farmer's best interest may or may not coincide with that of the community. If the high price is caused by a high level of demand in the market for new potatoes, then the farmer would be justified by any criterion in trying to meet that demand. Only if a large number of growers decided to harvest prematurely to secure the benefit of the high price would the community interest suffer: there would be an early glut on the market, followed by a shortage of mature crop later. Prices would drop for a time and then rise sharply; growers who had harvested early would not get the benefit since with the reversal of the prices their crop would fetch less than they expected, whilst by the time prices recovered they would have none left to sell and the balance would have to be imported. As only a very limited number of growers are in a location where favourable climate and relief allow them to make the choice, this particular difficulty is unlikely to arise; but in a year when the main-crop was likely to be poor (perhaps because of the weather), some form of inducement to growers to maximize yields rather than profits might bring about a net improvement in social benefits if the need for imports was reduced at the expense of, say, a fortnight's delay in having new potatoes on the lunch table. There could be a case for using subsidy policy or marketing boards to remove some of the unknowns from selling farm produce, as well as for employing the National Agricultural Advisory Service in the capacity of a co-ordinator of production, if national yields could be improved or the balance of staple crops better tailored to the needs of consumers as a result. Apart from specialized growers, who tend to find their own markets, there is also a case for aiding the type of farming best suited to particular regional locations and their associated climatic and edaphic conditions. Though there is always room for advice, the farmer himself is likely to have the most detailed local knowledge, and to transfe- the final judgment away from him would be a rash course. Selective subsidies adjust the balance of advantage without ousting this final choice, though they can vary from brilliant success to ill-judged distortion, or they can be merely ineffective. First and foremost, it is the intention which must be clear.

This leads to a point made by Stringer (Chapter 8, p. 87), that a farmer will not consider information or advice valuable if it is freely available to everyone. Whilst few would go so far as to ignore short-term weather forecasts, this attitude might make a co-ordinating role difficult if it was widely believed that the best way to make a profit was precisely to be out of line with the majority. To react rationally to a market situation is not always easy; but to assess total benefits and costs and then adjust hypothetical future prices so as to maximize net benefits is a very complex operation. Whether planning on a gigantic, perhaps national, scale would increase or decrease the physical dangers and marketing hazards to crops and stock is hard to determine. Simpson, Hales, and Fletcher (1963) describe an interesting experiment where linear programming was used to choose a plan of cultivation for some C.W.S. (Co-operative Wholesale Society) greenhouses. On this scale they were still able in their assumptions to take market prices as given. If there were to be a National Agricultural Advisory Service welfare plan, then this assumption would no longer hold and a monopoly supply curve would have to be used in the price analysis. Immunity from weather conditions was also assumed, so one might hope to use their data as a basis for comparative costs and yields between indoor and outdoor growing. Unfortunately, the winter of January–February 1963 was so cold that greenhouse temperatures could not be maintained, whilst the timing and yields of crops nationally differed radically from expectations and hence from the forecast prices. Luckily, both from the community's and growers' points of view, the recommended changes had not been implemented and the crop of tomatoes, which were scarce, had not given way to cucumbers, which were relatively plentiful.

Nevertheless, it is techniques of this type which would have to be used if agricultural planning were ever applied on a wider scale. Prices in this event could no longer be considered as immune from the effects of the plan, unless through the support policy; and weather hazards too, ironically, could acquire an entirely new role. A more systematic analysis of weather records and risks, and a wider application of the more sophisticated forms of weather protection can cheerfully be advocated before this stage is reached, since the welfare effects of knowledge and increased real assets are not in question.

Mention has also been made of labour availability to farmers. Taggart (Chapter 5, pp. 45–50) was concerned not so much with "what labour has the farmer got" but rather with "what use can he make of it". McFarquhar (1961a) took the view that if the farmer works ahead as far as he can, after giving priority to urgent work, he should have flexibility in hand to cover sickness or accidents. Larger businesses employing several hundreds of people can deal with absence by allowing for a percentage of staff to be away at any given time, which works, barring epidemics, as long as the individual operator's work is not so specialized that he cannot be replaced. On a farm there might be more trouble in replacing a labourer who was suffering from a prolonged illness, depending on the location. No mention is made of possible inability of workers to reach the farm in bad weather; perhaps it is assumed that in such severe conditions little work can usefully be undertaken anyway.

Labour availability in either sense does not seem to be a major problem in forestry. The objectives of the Forestry Commission are, as nearly as possible, synonymous with the public interest, and the effectiveness of the various steps described in Chapter 7 (pp. 67–82) is their own recommendation: they do not give rise to divergences between commercial interests and social benefits. Blown timber sold cheaply may benefit individual buyers, but there is no obvious reason why one customer should be favoured at the expense either of another or of the taxpayer. We can all rejoice that modernization of the railways has among other advantages reduced the number of forest fires, though the abolition of steam traction would have taken longer but for the complete elimination of many services in afforested country, which has certainly imposed social costs on some people.

Delays and difficulties in travelling to work, however, are the main effects of severe weather on manufacturing industry, as illustrated in

Chapter 8, pp. 88–90. Although there is still no theoretical divergence of interest between the local authorities and the public-transport operators on the one hand and the community they serve on the other, both manufacturers and individual members of the public can suffer losses which are not recoverable from any source, due partly to weather conditions, but often provoked or aggravated by difficulties of communication. It is true that the railway management, like the bus operators, generally spare no effort to keep scheduled services running in adverse weather (sometimes a modified service specially designed for foggy conditions). This may be due to an awareness not so much of potential losses to industry as of the danger of losing goodwill through failure to maintain the service. In any case neither of these notional cases could be accurately evaluated in the heat (or rather the chill) of the moment. No separate figures for costs of fighting adverse weather appear in the railway accounts; although some snow clearing can be performed by platelayers who cannot continue their normal work, much of it calls for motive power and special equipment. More people than usual take public transport to work in bad weather, but it is unlikely that the extra revenue from these standby users would meet the additional costs unless, as operators always fondly hope, they happen to find the service unexpectedly convenient and decide to patronize it more often in the future. In adverse conditions the risk is greater than usual that exactly the opposite will occur, and potential regular clients will experience delays, discomfort, and overcrowding which, though untypical, may dissuade them from using that particular form of public transport again.

The moral of the Birmingham experience (see pp. 88–90) is that people cannot make rational decisions unless they are in possession of accurate, comprehensive, and up-to-the-minute information. If this is to be provided by local radio, well and good: but the co-ordination of weather reports with local information on road conditions and transport services is essential if social costs are to be minimized. As important as the effective dissemination of these reports is the need to bluff or cajole the public into acting on the advice given.

Even if the cost of the time of road-menders and other employees (who were available because they could not continue with their normal work) has been excluded from the £55,000 spent by Birmingham Corporation on 4 days' road clearing, the sum is still small beer compared to the social cost of the likely delays if no clearing had been attempted, or, indeed, of the cost of the delays which did occur in spite of it. The true social return to be compared with this expenditure is the difference between these last two items. Some figures which have been received from Montreal (Department of Finance, City of Montreal, 1968)—which has about the same population as the west Midlands conurbation, but twice that of Birmingham alone—show that the average cost of removing snow there over the period from November 1958 to April 1968 exceeded $7 million at 1958 prices. It is possible that this includes all roads; whereas only the main thoroughfares were cleared in Birmingham. About $4·8 million of the total behaves like a fixed cost over the period, partly because over half the city area is let out to contractors for snow clearing at a fixed price; but this leaves a variable cost of some $27,000 per inch of snow in addition, which on its own is about double the *total* costs per inch quoted in the Birmingham example. The average annual snowfall in Montreal over this period was about 90 in. which also makes the use of contractors seem logical enough: the chance of a winter without snow is less than 1/100,000 on the basis of the data provided. The mean winter temperatures varied between −1° and −4°C.

The absence of net social benefit from insurance may be more obvious than it was in farming if it is postulated that local authorities and railways might choose to insure against snowfalls instead of investing the money in equipment to deal with them. The compensation received would be no substitute for getting on with the job. Insurance of this kind is also of little help to any large operator, since he can achieve the spreading of risks within his own organization. London Transport is bound to sustain accidental damage to a few buses each year, but the cost of repairing these is well below that of comprehensive premiums for the whole fleet. The laundry carries the risk of ruining clothes and pays compensation from the profits on those it cleans successfully; the G.P.O. pays compensation on the handful of registered packets that are lost from the registration fees for the thousands which arrive safely. Insurance firms exist to cover abnormal risks on the one hand, and to spread the burdens of possible loss on major possessions like a house and furniture, or a car, which may represent too large a proportion of an individual's assets for him to afford to lose. Because the probability of loss is low, the firms are almost certain to be meeting claims from only a few of their clients at any one time.

This is the basis of insurance against flood damage, a form of weather-derived devastation which has been news once more in 1968. Again, the maximum social benefit accrues from flood prevention, by means perhaps of dykes and adequate drainage. However, it is prudent for the individual to insure possessions against weather hazards as it is against fire, since the rate of risk is low but major damage could be cripplingly expensive. To cover for flood, frost, and gales costs little, if anything, over and above the premiums for fire and theft, though quite a number of house owners do not trouble with it. It has been suggested that flood damage should be made good from central government funds; but this penalizes the family who prudently choose or site their house 150 ft above the nearest water course, in a position sheltered from wind, and have to pay a little more in consequence, since they would also be required to subsidize the less prudent through taxation. Moreover, unscrupulous builders or local authorities would be encouraged to use sites where they knew quite well there was a flood risk. It is a simple matter for an actuary to compute the level of risk, if any, for a particular site on the evidence of past records, and to adjust premiums accordingly where necessary. If government intervention is needed at all, it is to induce insurers to make the wording of their policies intelligible. Local authorities could sponsor flood protection schemes in the collective interests of residents, making provision for the individual who could not have acted alone to contribute through his rates or otherwise.

The community gains from spending resources on protection rather than on insurance, but stands to lose if the cost of protection or insurance exceeds the long-term value of the assets protected. The critical threshold when measurable damage is caused can be defined for each type of hazard, together with the relationship between further damage and worsening conditions. In conjunction with this, a probability distribution of incidence of the hazard can be constructed from past weather records. The expected cost of damage at each level can be multiplied by the probability that the same level will be reached to give a net expectation of loss. Beyond a certain point the probabilities of such extreme conditions decrease more rapidly than the potential damage rises, so the net expected loss begins to decline. On the same basis as the cost/loss ratio described in Chapter 8 (p. 84), the annual expenditure on precautions should not exceed the highest figure reached before this decline begins. Alternatively, the probabilities can be treated as frequencies over the life of the protective assets (irrigation equipment, barns for wintering stock, or whatever it may be). If the expected life of these is 20 years, and the probability of losing say £1000 in any year is 0·25, but the probability of losing £1500 is 0·1, then the expected loss over the 20 years is five occasions at £1000, of which two would amount to £1500, giving a total of £6000. The principles of discounted cash flow must be applied to this loss since its impact would not, on average, be immediate. Strictly speaking, it should be spread evenly over the period, for want of any better assumption, but for simplicity it can be viewed as occurring 10 years hence. If the current cost of borrowing was 7 per cent, its present value would be

$$\frac{\pounds 6000}{(1 \cdot 07)^{10}} = \pounds 3050,$$

which is therefore the maximum to contemplate spending on protection in this example.

The only special difficulties about constructing probability distributions for weather hazards are that (a) the incidence of some of these may not be normally distributed, which would make the calculations less reliable unless another distribution could be found which gave a good fit, and (b) the various phenomena are not necessarily mathematically independent. This causes difficulties if it is necessary to calculate the risk of two or more occurring together because the resulting damage is then on a different scale from the sum of the several items separately. Examples of non-independence are that frost is more likely in calm, anticyclonic weather in autumn or winter than when there is good cloud cover and strong winds; frosty conditions may be associated with fog; floods are apt to occur in the rapid thaw after a heavy snowfall, and so forth. On the other hand such associations are not invariably present. In the present state of knowledge, the only safe way to construct joint distributions may be from records which show them occurring together; but data from the smaller and more dispersed meteorological stations, which is essential to achieve regional preductions, do not always cover a long enough time-period in sufficient detail to give usable results for many of the possible combinations.

ACKNOWLEDGEMENTS

I am grateful to Mr. W. Dyfri Jones of the Department of Agricultural Economics, University College of Wales, Aberystwyth, and to the staff of the Rural Science Library, University College of Wales, Aberystwyth, for helpful advice on some of the reference material consulted.

REFERENCES

CAMM, B. M. (1963) *Risk in Vegetable Production on a Fen Farm*, University of Cambridge School of Agriculture publication, pp. 89–98.

DEPARTMENT OF FINANCE, CITY OF MONTREAL (1968) Private communication.

DUNFORD, W. J. (1961) *Uncertainty and the Farmer*, Univ. of Bristol, Agricultural Report, No. 126.

LANGLEY, J. A. (1966) *Risk, Uncertainty, and Instability of Incomes in Agriculture*, University of Exeter, Dept. of Economics (Agricultural), Occasional Paper 2.

MCFARQUHAR, A. M. M. (1961a) The practical use of linear programming in farm planning, *The Farm Economist*, vol. IX, No. 10.

MCFARQUHAR, A. M. M. (1961b) Rural decision-making and risk in farm planning, *Journal of Agricultural Economics*, vol. XIV, No. 4.

SIMPSON, I. G., HALES, A. W. and FLETCHER, A. (1963) Linear programming and uncertain prices in horticulture, *Journal of Agricultural Economics*, December 1963.

Author Index

Page numbers in italics refer to figures thereon

Bodlaender, K. B. A. 53, 54, 65
Borah, M. N. 52, 53, 65
Brier, G. W. 84, 86, 91
Brooke, M. D. 97, 106

Camm, B. M. 108, 118
Chambers, R. 7, 9, 93, 106
Charlton, R. 14, 16
Charters, H. 68, 71, *72*, 81
Clarke, J. A. 100, 106
Croxall, H. E. 14, 16

Davies, J. W. 40, 43
Davis, L. S. 68, 69, 71, 74, 82
Doll, J. P. 51, *53*, 54, 65
Dorfman, R. 75, 82
Duckham, A N. 5, 9, 11, 13, 16, 104, 106
Dunford, W. J. 109, 119

Edlin, H. L. 6, 9
Edwards, R. S. 100, 106
Epstein, E. S. 86, 91

Finney, J. B. 14, 16
Fletcher, A. 113, 119
Fraser, A. I. 75, *76*, 82

Gardiner, J. B. H. 82
Gloyne, R. W. 28, 43, 94, 96, 106
Golding, E. W. 73, *73*, 82
Gregory, L. E. 53, 54, 65
Griffiths, D. J. 99, 106
Gringorten, I. J. 84, 91

Hales, A. W. 113, 119
Headford, D. W. R. 52, 65
Hogg, W. H. 6, 9, 27–44, 94, 95, 96, 99, 100, 101, 106
Hurst, G. W. 7, 9

Ivins, J. D. 65

Joce, E. H. B. 95, 96, 106
Johnson, G. 11, 16
Jones, L. R. 54, 65
Jones, W. D. J. 118

Kolb, L. L. 86, 91

Lamb, H. H. 101, 106
Langley, J. A. 108, 109, 119
Lave, L. B. 86, 91
Letnes, A. 52, 65

Mason, B. J. 7, 9
McFarquhar, A. M. M. 108, 114, 119
McQuigg, J. D. 13, 16, 51, *53*, 54, 65
Michael, D. T. 6, 9
Milnthorpe, F. L. 52, 53, 65
Morgan, J. P. (Miss) 99, 106
Murphy, A. H. 85, 86, 91

Penman, H. L. 32, 43, 52, 65
Pyatt, D. G. 76, 82

Radley, R. W. 59, 65
Rapp, R. R. 86, 91
Rouse, G. D. 81, 82
Rubra, G. N. 97, 98, 99, 100, 106, 107–19
Rutter, N. 103, 106

Samuelson, P. 74, 82
Shellard, H. C. 39, 43
Shorr, B. 85, 91
Simpson, I. G. 113, 119
Smith, C. V. 17, 95, 96, 106
Smith, L. P. 5, 9, 14, 28, 43, 98
Solow, R. M. 74, 82
Stansfield, J. M. 11–16, 93, 94, 106
Stodhart, A. H. 73, *73*, 82
Stringer, E. T. 83–92, 96, 97, 98, 104, 106, 113

Taggart, W. J. 45–50, 95, 96, 97, 99, 106, 114
Taylor, J. A. 5–10, 13, 16, 93–106
Theophilus, D. 55, *56*
Thompson, J. C. 84, 91
Timmis, A. (Mrs.) *55*
Toosey, R. D. 52, 65
Toulson, G. A. 94, 106
Tyler, G. J. 49, 50
Tyrrell, J. G. 45, 51–66, 96, 112

Wardle, P. A. 6, 67–82, 94, 95, 96, 97, 98, 99, 100, 106
Went, F. W. 54, 65

Subject Index

Page numbers in italics refer to figures thereon

Aberystwyth, University College of Wales 9, 106
  University College of Wales, Rural Science Library 118
Agriculture 3, 11, 16, 83–92, 102, 106
  Agricultural Land Service 106
  agricultural productivity 101
Animal diseases 8

Benefit/cost ratio 7
  see Cost/benefit analysis
Berkshire Downs 14
Biosphere 1
Birmingham 88, 89, 90, 104, 115, 116
  Birmingham Corporation 89
  Birmingham Evening Mail 88, 91
  Birmingham Post 88, 91
  University of 83, 106
Bristol *29*, 31, *32*, 106
  Filton 31
  Long Ashton 28, *29*, *32*, 43
  Westbury-on-Trym 106
British Electrical Research Association 82
British Insurance Association 7
British Rail 89, 90
Broad Haven 56

California 65, 69, 71, 82, 86
  Bel Aire, Los Angeles 67
  San Joaquin Valley 86
  State of California Resources Agency 82
  University of California, Los Angeles 65
Cambridge 196
Castlemartin *64*
Central Electricity Generating Board 8
  reserve capacity cost 8
Central heating 8
Chiltern Scarp 5
Chronica Botanica Co., Waltham, Massachusetts, U. S. A. 65
Climate 1, 49, 101, 102, 106, 112
  climatic factors 9
  climatic trends 101, 105
  climatology 90
  local climates 48
Community benefit *110*
Computer 103
Contract farming 93
Co-operative Wholesale Society 113
Cornwall 96
Cost/benefit analysis 5, 97, 104, 109
  individual 97
  national 97
  "social costs" 97
  studies 83–93
Cost/loss ratio 84, 85, 86, 91, 117

Dale *62*, *64*
  Dale Fort 61
Decision-maker 85, 88, 98
Decision-making 1, 3, 12, 13, 14, 16, 18, 84, 86, 87, 90, 93, 96, 98, 101, 104
  day-to-day operational decisions 11
  operational decisions 13
Deficiency payments 98
Degree-day 27–45, *40*, *42*
Disaster funds 2
  national 105
"Discounted cash flow" 100
Dorset 7, 93

Drought 102
Dry spells 21, *21*
  dry-day *23*

Economics *12*, 83, 90, 105
  economic activity 5
  economic systems 1
Edgbaston Observatory 83, 87, 88
Edinburgh 45, 46, 106
  Edinburgh School of Agriculture, University of Edinburgh 106
England, east Coast 103
  Britain 11, 67, 73
  south 82
  south-east 2, 3, 15, 27, 102, 104
  south-west 2, 27, 40, 41, 43
Entrepreneur 101–2, 103, 105
Environment 11, *12*, 108
Environmental factor 81
Environmental sciences 1
Europe 67
  the Continent 7
Experimental Husbandry Farms 14
  High Mowthorpe E.H.F. 14, 16

Farm Management Surveys 12
Farnham, Surrey 106
Filton *see* Bristol
Fire 67–68, *68*, 69–75, 80–81, 95, 97, 109, 114, 116–17
  losses *68*, *69*, *70*, *72*
  protection 80, 81, 95, 97
  risk 73–75, 95
  starts *72*
Fishguard 58
Floods 2, 3, 102, 103, 107, 116, 117, 118
  East Coast floods of 1953 103
  flood protection 117
  in south-east England, September 1968 2, 3
  in south-west England, July 1968 2
Foot-and-mouth disease 6, 7, 9
Forestry Commission 6, 68, *68*, *69*, 80, 81, 82, 95, 97, 100, 114
  Research Station, Alice Holt Lodge, Wrecclesham, Farnham, Surrey 106
Frost 5, 6, 27–43, *32*, 52, 59, *60*, 94, 99, 102–3, 107, 117–18
  frequency *29*
  killing frost 94
  protection 15, 94
  risk *31*, 94
  survey *30*

Gales 6, 67, 102
  1953 gale 71, 98
  January 1968, Central Scotland 6, 67, 71
General Post Office 116
Germany 67, 98
Glasgow 6
Gloucestershire 41
Growing season 32, 34

Haverfordwest 55, 59
H.M.S.O. 65, 106
Horticulture 27
Hurricane 2, 6

Ireland 96
Irrigation 6, 15, 18, 27–44, *34*, *35*, *61*, *62*, 61–63, 65, 94, 96, 98–100, 102
  Atlas 43, 94 (*see under* Hogg)
  -need *34*, *35*, *36*, *37*, *38*, *39*, 43, 61, 94, 99, 106
  of blackcurrants 100

Labour 11, *12*, 13, 45, 48–50, 97
  agricultural 45–50
  availability 48, *49*, 49–50, 114
  force 3, 16, 25
  marginal value of *50*
  optimal deployment of 97
  resources 3
Laissez-faire 105
Lancashire, mossland 5, 94
  south-west Lancashire 93
Leeds, Lawnswood 106
Lexington, Kentucky, University of Kentucky 16
Little Haven *64*
London Transport 116
Long Ashton *see* Bristol

Madrid 82
Management 11, 12, *12*, 13, 27, 74, 95, 106, 108, 109
farm 9, 11–16, 96
forest 6
Marloes Peninsula 63
Marloes 63, *64*
Mathry 56
Meteorological Office 1, 7, 19, 26, 43, 95, 103, 106
Meteorology 1, 9, 83, 84, 90
conditions 7
factors 9
services 7
Midlands 83, 88, 89, 98
East Midlands 22, *23*, 24, *24*
West Midlands 40, 43, 88, 106, 116
Ministry of Agriculture, Fisheries & Food 32, 40, 43, *55*, 65, 106
Brooklands Ave., Cambridge 106
Crown Buildings, Plas Crug, Aberystwyth 106
Great Westminster House, London 106
Woodthorne, Wolverhampton 106
Missouri, University of Missouri 65
Molinia (*Molinia caerulea*) 95
Montreal, Department of Finance 116, 118

National Agricultural Advisory Service 13, 14, 20, 22, 55, *56*, 100, 106, 113
Lawnswood, Leeds 106
Trawscoed, Cardiganshire 106
Westbury-on-Trym, Bristol 106
(W. Midlands Region), Woodthorne, Wolverhampton 106
National Institute of Agricultural Engineering 94
Natural Environment Research Council 2
Natural Resources Committee 61, 65
Newcastleton Forest, Roxburghshire 82
Nottingham *21*
University of 65, 106

Oswestry 7
Oxford 34, *33*

Pembrokeshire 59, 61, *64*, 96
north Pembrokeshire 56
Pembrokeshire coast 39, *39*, 51, 54
south Pembrokeshire 56
south-west Pembrokeshire 65
west Pembrokeshire *55*, *56*, *57*, *58*, *60*, *61*
Penzance, Cornwall 40, *40*
Potato 5–6, 8, 15, 27, 39, *46*, 48, *49*, 51–54, *53*, *56*, 59, 61–63, 65, 94, 96, 112–13
blight 8, 96
early potatoes 15, 27, 39, *46*, 48, 51–66, *55*, *56*, 96, 112
var. *Home Guard* *57*, 63, *64*
first lifting dates *57*
prices *54*
var. *Red Craig's Royal* 15
Prescelly Mountains 59
Probability forecast 83–93
Productivity
agricultural 101
variations due to 94

Rainfall 2, 13, 15, 17–22, *23*, 24, 27, 32–34, *33*, 46, *53*, 59, 61, *62*, 73, 80, 82, 86, 102–3, 106–7
Rainford 94
Reading 15, 106
University of *12*, 14
University Farm, Sonning-on-Thames 106

Scotland 2, 6, 67, 95, 97, 99, 106
Shap Fell 7
Sheep 6, 9, 15
Shrewsbury 37, *39*
Sitka spruce (*Picea sitchensis*) 77, *78*, 82
Snowstorms 88, 89, 90, 102, 104, 107, 115, 116, 118
snows 3, 7
Soil moisture *53*, 65
deficit 19, 32, *62*
Somerset 7, 29, *30*, *31*, 31, 93, 94
St. Bride's 63
St. David's 58, *64*
St. David's Peninsula 55
Subsidies 98, 99, 105
Sugar beet virus 8
Sutton Bonington 106

Systems analysis 84

Tasmania 67
Tenby 56
Tornado 5
Trough (of low pressure) 2

United States of America 8, 13, 67

Vale of Aylesbury 5

Wageningen 65
Wales 6, 27, 43, 94
  south Wales 41
  south-west Wales 51–56
  west Wales 106
Weather 1, 3, 5–10, 13–16, 17–26, 40, 45–46, *47*, 50, 51–52, 53, 54, 56, 57, 59, 63, 65, 67–73, 83–84, 87–91, 93, 95–96, 102, 104, 106, 107–9, *110*, 112–15, 118
  "applied weather forecasts" 96
  cost of British 5–9
  data 5, 16, 18
  effect on labour availability 97
  factors 5, 6, 7, 8 27, 45–50, *53*
  forecast 2, 13, 86, 90, 91, 96, 103
  losses to industries due to 7
  probabilities 3, 94, 99, 100, 101
  protection 15, 114
  protection devices 5, 8, 9, 102
  sensitive 3, 5, 11, 17, 59, 63, 64, 104, 108
  variation in milk yields due to 93
Weather-costing 5
Weather hazards 3, 6, 8, 9, 11, 15, 16, 81, 87, 90, 93, 94, 96–106, 108–9, 114, 116, 118
  variability of 102–3
  vulnerability to 102–3
Weather insurance
  for farmers, builders and civil engineers 98
  for fruit-growers 98
  premiums 105
Weather-proofing 2, 3, 14, 98, 101, 103, 104, 105
Westmorland 7
Wheat, var. *Capelle* 14
  winter wheat 16
Wiltshire 6, 7, 93
Wind-blow 67, 95
Wind-machines 94, 98
Wind-throw 67, 68, *68*, 69, *70*, *71*, 71–76, 79–82, 94–95, 98
Wolverhampton, Woodthorne 106
Work-day 17–26, *24*, 97
  outdoor work *47*, *48*
  work criteria *23*, 24